FLORA OF TROPICAL EAST AFRICA

MYRTACEAE

B. Verdcourt

Dedicated to the memory of Gerda Jane Hillegonda Am
(5 Jan. 1913–10 Feb. 1985)*

Trees, shrubs or occasionally pyrophytic subshrubs with massive rootstocks; usually evergreen; pith with internal phloem. Leaves simple, predominantly opposite, often coriaceous, mostly entire, glandular-punctate; stipules absent or very reduced. Flowers mostly regular, hermaphrodite or unisexual, solitary or in simple to complex inflorescences; bracteoles often present. Hypanthium ("calyx-tube") ± adnate to the ovary; lobes 3–6(–10), imbricate, valvate, splitting irregularly or forming an operculum. Petals 4–5(–6) or rarely absent, included on the margin of the disc lining the calyx-tube, imbricate or forming an operculum. Stamens numerous, rarely only 4, 5 or 10, included on the disc margin in 1 or more rows, straight, inflexed or twice folded in bud; filaments free or connate at the base into a short tube or in 4–5 bundles opposite the petals; anthers small, 2-locular, opening by slits or less often by apical pores; connective sometimes tipped by a gland. Ovary inferior or ± superior, (1–)2–5(–16)-locular with axile or rarely parietal placentation; ovules (1–)2–many, anatropous to campylotropous; style terminal (absent in one genus). Fruit a berry or capsule (less often a drupe or nut), (1–)few(–many)-seeded, indehiscent or loculicidally dehiscent. Seeds without or with very little endosperm; embryo straight, incurved, circular or spiral.

A large family with about 127 genera and 3850 species in the tropics to warm temperate regions of both Old and New Worlds; particularly well represented in Australia. The two native genera with about 21 species are of extreme taxonomic difficulty; in addition a large number of species is cultivated including between 100–150 species of *Eucalyptus*. Genera which contain widely cultivated species or species which have become naturalised are dealt with below. All are included in the key to genera. Cultivated species belonging to genera dealt with in the main text will be found there.

1. Fruit a capsule, loculicidally dehiscent at the apex with the number of apical valves equalling the number of loculi · 2
   Fruit a berry · 12
2. Sepals and petals fused into one or two opercula falling at anthesis (*Eucalypteae*) · · · · · · · · · · · · **Eucalyptus** (including *Corymbia*) (p. 15)
   Sepals and petals free · 3
3. Anthers basifixed (*Calothamneae*) · · · · · · · · · · · · · **Calothamnus** (p. 3)
   Anthers dorsifixed and versatile · 4

* see Taxon 34: 570 & 763 (1985)

1

4. Flowers solitary (mostly sessile) in leaf-axils or in
   axils of bracts, often appearing spike-like with
   leafy axis frequently projecting beyond (note a
   few *Eucalyptus* species have solitary flowers so if
   no buds available and only fruits this should be
   borne in mind) · 5
   Flowers in various inflorescences, not solitary · · · · · · · · · · · · · · · 7
5. Stamens partly united into bundles opposite the
   petals (*Melaleuceae*) · · · · · · · · · · · · · · · · ·   **Melaleuca** (p. 7)
   Stamens free or only fused at base (in 2 species of
   Callistemon) (*Leptospermeae*) · 6
6. Stamens not longer than petals · · · · · · · · · · · ·   **Leptospermum** (p. 3)
   Stamens long-exserted · · · · · · · · · · · · · · · · · ·   **Callistemon** (p. 5)
7. Flowers more or less partly connate by lower part
   of the calyx into globose pedunculate heads;
   calyx-lobes and petals separate (*Syncarpieae*)
   (some *Eucalyptus* species have connate flowers
   and a fruiting specimen with no available buds
   might key here) ·   **Syncarpia** (p. 11)
   Flowers free but sometimes in tight heads · 8
8. Stamens ± 20; flowers in sessile globular axillary
   heads ·   **Agonis** (p. 3)
   Stamens numerous or in bundles · 9
9. Calyx, inflorescence-axes and usually the
   undersides of the leaves grey velvety tomentose;
   leaves elliptic, opposite, 2.5–10 × 1–5.2 cm;
   stamens 3–3.8 cm long · · · · · · · · · · · · · · · ·   **Metrosideros** (p. 13)
   Without above characters combined · 10
10. Stamens in 5 bundles (*Tristanieae*) · · · · · · · · · · ·   **Lophostemon** (p. 11)
    Stamens numerous in several series, free or in very
    obscure bundles (*Eucalypteae*) · 11
11. Petals distinct ·   **Angophora** (p. 11)
    Petals and sepals united into a circumscissile
    operculum (*Eucalyptus* specimens in flower
    without any buds available will key here but no
    calyx-lobes or petals will be visible) · · · · · · · · ·   **Eucalyptus** (p. 15)
12. Calyx-lobes, free part of calyx-tube and
    androecium circumscissile after flowering so
    that fruit has a flat circular scar (species
    cultivated in East Africa has leaves very acutely
    narrowly acuminate and inflorescences sessile in
    leaf-axils) ·   **Myrciaria** (p. 15)
    Calyx-lobes or at least free part of calyx-tube
    persistent on the fruit · 13
13. Stamens ± straight in bud; seeds with endosperm;
    inflorescences with few large flowers · · · · · · · ·   **Acca** *(Feijoa)* (p. 13)
    Stamens inflexed in bud; seeds without endosperm · · · · · · · · · · · · · · · 14
14. Calyx limb closed in bud, deeply divided when
    flowering; ovary 4–5-locular; seeds mostly very
    numerous; stigma usually capitate or peltate · ·   **1. Psidium** (p. 52)
    Calyx 4–5-lobed in bud, not more deeply divided
    when flowering; ovary 1–2(–3)-locular; seeds
    1–4(–17) ; stigma filiform to slightly capitate · · · · · · · · · · · · · · · · · 15

15. Ovules 1–7, pendulous from apex of each locule;
    inflorescences many-flowered; cotyledons small   **Pimenta** (p. 13)
    Ovules 1–numerous, not pendulous, often
    numerous on axile placentae; inflorescences
    1–many-flowered; cotyledons small or large, if
    small then flowers solitary · 16
16. Embryo curved, circular or spiral with small
    cotyledons; placentation in upper part of ovary
    appearing parietal; cultivated plant with solitary
    flowers; leaves small, rigid, lanceolate, ending in
    a stiff point ·   **Myrtus** (p. 13)
    Embryo thick and fleshy with large cotyledons
    much larger than the hypocotyl (connate or
    completely fused in *Eugenia*); placentation not
    as above; inflorescences 1–many-flowered or if
    flowers solitary then leaves not as above · · · · · · · · · · · · · · · · · · · 17
17. Calyx-tube not or scarcely prolonged above the
    ovary; stamens included on a disc surrounding
    the style; flowers few in axillary inflorescences ·   2. **Eugenia** (p. 55)
    Calyx-tube generally prolonged above the ovary;
    stamens on the border of the floral tube; flowers
    more numerous in terminal inflorescences · · ·   3. **Syzygium** (p. 67)

CALOTHAMNUS *Labill.*

Jex-Blake, Gard. E. Afr. ed. 4: 106 (1957) mentions **Calothamnus quadrifidus** *R. Br.* and it has presumably been cultivated in Kenya but I have seen no material. It is certainly grown in Zimbabwe. Ericoid shrub or tree to 2.5 m with corky bark; leaves linear, 2–2.5 cm long, densely spirally arranged; flowers sessile forming one-sided inflorescences at base of branchlets (a "one-sided bottle brush"); staminal bundles and style crimson, ± 3 cm long.

AGONIS *(DC.) Sweet*

Agonis flexuosa *Schauer* has been grown in Kenya (Kiambu District: Lari Arboretum, 12 Mar. 1953, *Darling* 48 & Limuru, 11 Feb. 1955, *Brown in EA* 10629). A native of Western Australia, it is a much branched tree to 9 m with narrow sharply acute lanceolate alternate leaves and sessile, very tight (but not fused) heads of white flowers.

LEPTOSPERMUM *J.R. Forst. & G. Forst.*

Several species of *Leptospermum* have been grown as ornaments and the names of some are not altogether clear. The genus has been monographed by J. Thompson in Telopea 3: 301–448 (1989).

1. Ovary and fruits 6–11-locular; leaves 5–8 mm wide · · · · · · · · · · · · · ·   1. *L. laevigatum*
   Ovary and fruits 5-locular; leaves 1.2–5.5 mm wide ·2
2. Leaves 2.2–5 cm long, 3–3.5 mm wide · · · · · · · · · · · · · · · · · ·   2. *L. petersonii*
   Leaves 1.2–1.5 cm long, 1.2–2.2 mm wide ·3
3. Leaves spreading ·   3. *L. continentale*
   Leaves ± erect ·   4. *L. juniperinum*

1. **L. laevigatum** *F. Muell.* The 'Australian Myrtle' (Jex-Blake, Gard. E. Afr. ed. 4: 118 (1957)) is widely grown in Kenya as an ornamental shrub (Nakuru District: Solai, Gilbert Walker's Farm, 21 July 1952, *Bally* 8230; Kiambu District: Lari Arboretum, 12 Mar. 1953, *Darling* 41). Shrub 4.5–6 m tall with small oblong elliptic or narrowly obovate leaves 1.5–3 cm long, 5–8 mm wide and solitary ± sessile axillary flowers with white petals; capsule 6–11-locular. Fig. 1/1–2, p. 4

FIG. 1. *LEPTOSPERMUM LAEVIGATUM* — **1**, habit, × ²/₃; **2**, fruit, × 2. *PIMENTA DIOICA* — **3**, habit, × ²/₃; **4**, flower, × 3. *MELALEUCA QUINQUENERVIA* — **5**, habit, × ²/₃; **6**, flower, × 3; **7**, fruit, × 3. 1–2 from *Bally* 8230 with some flower information from *H.C. Evans* s.n. from South Africa; 3–4 from *Greenway* 1625; 5–7 from *Burtt* 130. Drawn by Pat Halliday.

2. **L. petersonii** *F.M. Bailey* (*L. citratum* Challinor, Cheel & A.R. Penfold) is also widely grown in Kenya (Nakuru District: Solai, Gilbert Walker's Farm, 21 July 1952, *Bally* 8233; Kiambu District: Lari Arboretum, 12 Mar. 1952, *Darling* 56; Nairobi Arboretum, 17 Mar. 1952, *Williams Sangai* 375). A much branched tree 7.5–9 m tall with lemon-scented foliage; bark fissured longitudinally. Leaves lanceolate or almost linear, 2.2–5 cm long, 3–5.5 mm wide. Petals white. Capsule 5-locular.

3. **L. continentale** *J. Thomps.* (*L. scoparium* auctt. *non* J.R. Forst. & G. Forst., Jex-Blake, Gard. E. Afr. ed. 4: 118 (1957)) has been grown in Uganda (Mengo District: Entebbe Botanic Gardens) and in Kenya (Nakuru District: Solai, Gilbert Walker's Farm, 21 July 1952, *Bally* 8231; Kiambu District: Muguga, 5 Aug. 1963, *Greenway* 10884). Ericoid shrub about 2.5 m tall with spreading linear or linear-elliptic leaves 1.2 cm long, 2.2 mm wide. Petals white, pink or crimson. Fruit 5-locular.
    NOTE. Most material cultivated throughout the world has been called *L. scoparium*; the two are very close and intermediates occur. *L. scoparium* is keyed as having leaves over 3 mm wide and often more than 5 mm. *L. continentale* is a native of continental Australia and *L. scoparium* of New Zealand, Tasmania and a restricted area of Australia.

4. **L. juniperinum** *Sm.* A Ugandan specimen (Mengo District: Entebbe Botanic Gardens, Oct. 1952, *Dale* U805) with linear non-spreading leaves 1.5 cm long, 1.2 mm wide probably belongs to this species. Dale, Introd. Trees Uganda: 46 (1953) records *L. scoparium* from there.

CALLISTEMON R. Br.
Callistemon species ("bottle-brushes") are popular ornamentals and at least eight have been grown. C.M. Mitchem gives a key in The Plantsman 15: 29–41 (1993) and Lumley & Spencer deal with some problems in Muelleria 6: 411–415 (1988).

1. Filaments fused at base · 2
    Filaments free · 3
2. Branches weeping · 1. *C. viminalis*
    Branches not weeping · 2. *C. citrinus*
3. Leaves less than 4 mm wide · 4
    Leaves more than 4 mm wide · 5
4. Leaves usually more than 4 cm long · · · · · · · · · · · · · · · · · 3. *C. rigidus*
    Leaves less than 4 cm long · 4. *C. subulatus*
5. Filaments cream, yellow or green · 6
    Filaments pink to red · 7
6. Filaments yellow to green; rachis glabrous to sparsely pubescent; leaves
        narrowly elliptic · 5. *C. salignus*
    Filaments cream to pale yellow; rachis more densely pubescent; leaves
        broadly elliptic to oblanceolate · · · · · · · · · · · · · · · · · · · 6. *C. pallidus*
7. Leaves to 7.5 cm long with tuberculate oil-glands · · · · · · · · · · · · · 7. *C. rugulosus*
    Leaves over 7.5 cm long; oil-glands not tuberculate · · · · · · · · · · · · · · · · · · 8
8. Leaves narrowly oblanceolate, thick and rigid with abruptly tapering
        apex · 8. *C. phoeniceus*
    Leaves linear to narrowly elliptic, thin and flexible with gradually
        tapering apex and ridged margins · · · · · · · · · · · · · · · · · · · 3. *C. rigidus*
    (Some *C. citrinus* will also key here but have elliptic leaves without prominently ridged
        margins)

1. **C. viminalis** (*Gaertn.*) *Loudon* (= *Melaleuca viminalis* (Gaertn.) Byrnes; Byrnes (Austrobaileya 2: 261 (1986) puts this in *Melaleuca* because of the shortly joined stamens but does not mention *C. citrinus*). Widely cultivated in East Africa (Kenya. Trans-Nzoia District: Kitale Museum Nature Trail, 19 Sept. 1984, *Mungai* 204/84; Kiambu District: Muguga Arboretum, 16 Aug. 1975, *Dyson* 687 & Hort. Greenway, 18 Aug. 1963, *Greenway* 10887. Tanzania. Lushoto District: Lushoto, Lawns Hotel Garden, 8 Feb. 1969, *Ruffo* 195 & 30 Jan. 1981, *Mtui* 41; Morogoro District: Morogoro, Agric. Dept., 13 Oct. 1938, *Greenway* 5811. A number of other specimens probably belong here but the habit data are lacking – Tanzania. Lushoto District: Amani, 31 Oct. 1969, *Ngoundai* 425 & Mkuzi, Hort. Greenway, Dec. 1960, *Salimu* 8 & Mar. 1959, *Mgaza* 242 & Mazumbai, Hort. Tanner, 7 July 1966, *Semsei* 4058; Morogoro District: Morogoro, Nov. 1955, *Semsei* 2389; Iringa District: Mufindi, Kigogo Forest Reserve, 10 Oct. 1954, *Sangiwa* 46).

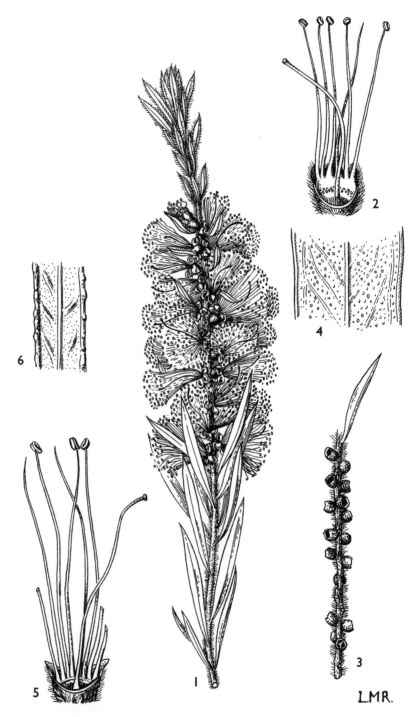

FIG. 2. *CALLISTEMON VIMINALIS* — **1**, flowering shoot, × ²/₃; **2**, opened flower, × 2; **3**, fruiting twig, × ²/₃; **4**, leaf surface, × 4. *C. RIGIDUS* — **5**, opened flower, × 2; **6**, leaf surface, × 4. 1–4 from *Selman & Jabri* 267; 5–6 from *Omar & Sahira* 363. Drawn by L. Mason Ripley. From Flora of Iraq 4(1), t. 68.

A small multi-stemmed tree about 6 m tall; branches and inflorescences pendulous. Stamens deep crimson; anthers greenish black not contrasting with dark stigma. Fig. 2/1–4.

I do not know how one tells *C. citrinus* and *C. viminalis* apart without habit data.

2. **C. citrinus** (*Curtis*) *Skeels* (*C. speciosus* auctt. *non* DC., *C. lanceolatus* Sweet) (Jex-Blake, Gard. E. Afr. ed. 4: 106 (1957)) has been recorded many times but most specimens seem to belong to the previous species (Kenya. Nairobi, National [Coryndon] Museum Grounds, near Desert Locust Control Building, 13 Mar. 1952, *Williams Sangai* 366; Tanzania. Iringa District: Mufindi, near streams in miombo woodland, 7 Oct. 1936, *McGregor* 14 (was this naturalized?)). Brenan (T.T.C.L.: 370 (1949)), records both *C. speciosus* and *C. lanceolatus*.
Small tree to 7.5 m with narrow leaves 0.5–2.5 cm wide; not weeping. Stamens red.

3. **C. rigidus** *R. Br.* occurs in the Usambaras (Tanzania. Lushoto District: Malindi, 20 Aug. 1932, *Geilinger* 1543 & Amani, 19 Feb. 1971, *Furuya* 210 & Kiumba, Plantation 1, 21 Oct. 1930, *Greenway* 2548) (Jex-Blake, Gard. E. Afr. ed. 4: 106 (1957)).
A shrub 2.4–4.5 m tall; filaments red; anthers dark red; stigma yellow. Dale records it for Uganda (Introd. Trees Uganda: 15 (1953); T.T.C.L.: 370 (1949)). Fig. 2/5–6.

4. **C. subulatus** *Cheel.* A specimen at first wrongly associated with the previous species appears to belong here (Kenya. Kiambu District: Muguga, E.A.A.F.R.O. Estate, 16 Aug. 1975, *Dyson* 686).
A stiff erect shrub 1.5 m tall. Stamens deep carmine, anthers dark plum-coloured contrasting with yellow stigma.

5. **C. salignus** *Sm.* is a well known cultivated species in East Africa (Jex-Blake, Gard. E. Afr. ed. 4: 106 (1957)) (Uganda. Mengo District: Kampala, Makerere University Hill, 10 Sept. 1971, *Lye* 6053; Kenya. Nairobi, Arboretum Block XXI, 7 Mar. 1952, *Dyson* 238; Kiambu District: Muguga, Hort. Greenway, 22 Sept. 1963, *Greenway* 10895 & Muguga, E.A.A.F.R.O. Estate, 16 Aug. 1975, *Dyson* 685).
Shrub or small tree 3–6 m tall. Young leaves silvery pubescent, later olive green. Petals greenish white. Stamens white, anthers yellow. Style apple green.

6. **C. pallidus** (*Bonpl.*) *DC.* A single specimen probably collected in the 1920s is likely to be this (Kenya. Nairobi, Karura Forest, *Battiscombe* 1006) but colour data is lacking. The tight inflorescence buds, about 2.5 cm long, have not been seen in Australian material.

7. **C. rugulosus** (*Link*) *DC.* (*C. coccineus* F. Muell.). (Kenya. Kiambu District: Muguga, E.A.A.F.R.O., 16 Aug. 1975, *Dyson* 688; Nairobi, Sept. 1954, *Wood* H 232/54).
Shrub to 3.5 m. Stamens red with conspicuous yellow-green anthers.

8. **C. phoeniceus** *Lindl.* (Jex-Blake, Gard. E. Afr. ed. 4: 106 (1957)) (Kenya. Nairobi Arboretum, 25 July 1952, *Williams Sangai* 494).
Much branched tree to 7.5 m with red flowers.

MELALEUCA *L.*
About 16 species of *Melaleuca* have been seen or recorded from East Africa. The genus has been partly revised by Byrnes in Austrobaileya 1: 65–76 (1984), 2: 131–146 (1985) & 2: 254–276 (1986). The *Melaleuca leucadendron* group was revised by S.T. Blake in Contr. Queensland Herb. No. 1, 114 pp (1968). Several species have been persistently misidentified.

1. Leaves 5–20 cm long, mostly over 1 cm wide · 2
   Leaves under 5 cm long · 3
2. Leaves 8.5–19 × 0.9–2.5 cm, 9–14 times as long as wide; spikes
   6–15 × 2.2–3 cm with glabrous rachis; petals with small elliptic
   but usually no linear glands ·   8. *M. leucadendron*
   Leaves 5–9 × 0.6–2.4 cm, (3–)4–6(–7) times as long as wide;
   spikes 4–8.5 × 2.5–3.5 cm with rachis almost glabrous or
   sparsely to densely hairy; petals with   linear and elliptic
   glands ·   9. *M. quinquenervia*
3. Leaves predominantly opposite · 4
   Leaves alternate or spirally arranged, a few sometimes appearing
   opposite · 9

4. Leaves narrowly elliptic with ± revolute edges, ± 2.7 × 0.8–1 cm; spikes 7.5 × 4.5–5 cm; stamens and style exceeding 2 cm · · · ·   2. *M. hypericifolia*

    Leaves much narrower, linear to linear-elliptic (save in a variety of *M. fulgens* with leaves up to ± 2 × 0.5 cm) · 5

5. Leaves linear, 1–6 cm × 1–2.5 mm; inflorescences few-flowered with well-spaced flower-pairs; petals membranous, 5 mm diameter; staminal bundles up to 1 cm long. Style thick, about 1 cm long with distinct mushroom-shaped stigma 1.5 mm wide; fruits globose, about 9 mm diameter, thick-walled with rather narrow opening ·   7. *M. radula*

    Without above characters combined · 6

6. Leaves narrowly oblong, ± 2 × 0.5 cm · · · · · · · · · · · · · · · · · ·   3. *M. fulgens* subsp. *steedmanii*

    Leaves linear to linear-elliptic, much narrower · 7

7. Staminal claw linear with short filaments; leaves linear, 10–32 × 1–3 mm; flowers white (layered papery bark) · · · · · · · · · · · ·   5. *M. linariifolia*

    Staminal claw narrowly oblong with long filaments; leaves 5–35 × 0.4–3 mm; flowers red or purple · 8

8. Leaves linear-elliptic, 5–15 × 1–3 mm · · · · · · · · · · · · · ·   4. *M. thymifolia*

    Leaves linear, 8–35 × 0.4–1.5 mm ·   3. *M. fulgens*

9. Flower buds subtended by sinuous bracts ± 1 cm long · · · · · · ·   1. *M. macronychia*

    Flower buds without such obvious bracts · 10

10. Leaves ovate-lanceolate, ± 5 × 1.7 mm (3–5-nerved; inflorescences ± 11 cm long, stamens white, perianth pink) · · · · · · · · · · · ·   12. *M. huegelii*

    Leaves longer · 11

11. Leaves very closely ± 20-nerved, 22 × 5 mm (bark very soft, in numerous papery layers; flowers white) · · · · · · · · · · · · · · ·   11. *M. styphelioides*

    Leaves with far fewer nerves · 12

12. Leaves ± oblanceolate, 10–40 × 3–9 mm, rounded at apex; flowers pink or purple-red in terminal heads 1.8–2.5 cm diameter usually soon overtopped by leafy axes; filaments 10–15 per bundle ·   16. *M. nesophila*

    Leaves not oblanceolate and without other characters combined; flowers white or yellowish · 13

13. Leaves not petiolate, linear-elliptic to linear-lanceolate, pubescent or glabrous, up to 7-nerved but often not visible, 5–20 × 1–2.5 mm, pungent ·   10. *M. bracteata*

    Leaves petiolate or if petiole obscure then leaves linear · · · · · · · · · · · · · · · · · · 14

14. Leaves linear-elliptic to linear-lanceolate, 4–13 × 1.1–1.5(–2.3) mm; branches often densely pubescent; staminal claws 0.5–1.8 mm long with 6–14 filaments (bark rough) · · · · · · · · · · · ·   14. *M. lanceolata*

    Leaves linear or, if linear-lanceolate the staminal claws ± 6 mm long · 15

15. Leaves more linear-lanceolate, not obviously gland-dotted, 12–14 × 1.5–2 mm; staminal claws 6 mm long with (15–)26–30 filaments (bark papery) ·   13. *M. preissiana*

    Leaves linear · 16

16. Leaves with obvious gland-dots; staminal claw with filaments arising from entire length ·   6. *M. alternifolia*

    Leaves without such obvious gland-dots; staminal claw with filaments arising only from apical 1–2 mm · · · · · · · · · · · · ·   15. *M. armillaris*

1. **M. macronychia** *Turcz.* (*M. longicoma* Benth.; Jex-Blake, Gard. E. Afr. ed. 4: 119 (1957))
Shrub to 3 m, glabrous save for very young shoots and inflorescence axis. Leaves spirally arranged, oblanceolate or narrowly oblong-elliptic (more broadly elliptic in one variety) 0.9–2.7 cm long, 4.5–9(–13) mm wide, acute, the venation rather obscure, gland-dotted beneath. Inflorescences lateral many-flowered spikes 2.3–4.7 cm long, up to 4 cm wide, the flower buds subtended by lanceolate acuminate bracts 8–10 mm long which soon fall. Petals 4–5 mm long. Stamen claws yellow, 0.8–1.7 cm long with 20–34 crimson-tipped filaments.

2. **M. hypericifolia** *Sm.* has been grown in Kenya (Kiambu District: Muguga, Hort. Greenway, 6 Sept. 1963, *Greenway* 10893 (Jex-Blake, Gard. E. Afr. ed. 4: 119 (1957)).

A much branched aromatic evergreen shrub to 2.4 m with ± horizontal branches, narrowly elliptic opposite ± revolute leaves 2.7 cm long, 0.8–1 cm wide. Spikes 7.5 cm long, 4.5–5 cm wide, flowers dark pink; anthers rich purple. Style and stigma green.

3. **M. fulgens** *R. Br.* The material so-called (Kenya. Nairobi Arboretum, 18 Feb. 1952, *Williams Sangai* 343 which bears a note = *Gardner* 1161 which must be a specimen at EA not duplicated at Kew) is none too good a match for this species and has a poorly developed inflorescence.

Shrub 0.4–2.6 m tall with opposite linear leaves 0.8–3.5 cm long, 0.4–1.5 mm wide and reddish spikes 3.5 cm long; staminal claws scarlet or deep pink, rarely white, 0.3–1.4 cm long with 22–80 filaments. Young fruits bear a persistent long style with a capitate stigma but not visible amongst the stamens in flowering state; possible separate male and female or hermaphrodite flowers.

NOTE: Further investigation into the sexuality of flowers is needed. Subsp. *steedmanii* (C. Gardner) Cowley has been recorded from Nairobi but I have not seen material since 1964. It has narrowly oblong leaves ± 2 × 0.5 cm and stamen bundles 1.5–2 cm long.

4. **M. thymifolia** *Sm.* is recorded as having been cultivated in Nairobi and Solai (Gilbert Walker's Farm, 21 July 1952, *Bally* 8229) and mentioned in Jex-Blake, Gard. E. Afr. ed. 4: 119 (1957).

Shrub usually only up to 1 m but recorded up to 6 m; usually with many branches from a woody tuber; bark corky, flaking; branchlets puberulous or quite glabrous. Leaves opposite or subopposite, lanceolate to narrowly elliptic, 0.5–1.5 cm long, 1–3 mm wide, acute, usually glabrous, 3-nerved but often only midrib evident, ± strongly gland-dotted beneath, the margins slightly to distinctly thickened. Inflorescences ± few-flowered dense axillary spikes usually on older wood, ± 1.5 cm long with flowers solitary within each bract. Petals pink to purple, 4–5 mm long. Staminal bundles pink to purple with claws 4–6 mm long and 40–60 filaments.

5. **M. linariifolia** *Sm.* has been cultivated in Nairobi (Nairobi Arboretum, 12 June 1962, *Stewart* H12/62).

Shrub or small tree to 10 m with layered papery soft bark; branchlets pubescent, soon glabrous. Leaves opposite or subopposite, linear-lanceolate, 1–3.2 cm long, 1–3 mm wide, acute, glabrous, 3-nerved with oil glands usually very distinct beneath. Inflorescences many-flowered usually open terminal or subterminal spikes with flowers solitary within each bract, usually opposite. Petals white, 2–3 mm long. Staminal bundles white with claws 0.4–1.5 cm long and 30–60 filaments.

6. **M. alternifolia** (*Maiden & Betche*) *Cheel.* Two specimens named *M. lateritia* A. Dietr. appear to be this species (Tanzania. Lushoto District: Amani, 4 Aug. 1970, *Furuya* 137 & Lushoto, State Lodge, 7 Oct. 1974, *Ruffo* 1031 & idem, 11 Mar. 1964, *Mgaza* 584).

Shrub or small tree 4–7 m tall with soft layered peeling papery bark; branchlets glabrescent. Leaves alternate, linear, 1–3.5 cm long, usually under 1 mm wide. Flowers white, solitary within each bract arranged in many-flowered spikes, terminal or from upper axils. Petals rounded ovate, 2 mm long and wide. Staminal bundles with claw 11 mm long and 30–35 filaments spread all over claw, those at base shortest.

7. **M. radula** *Lindl.* is mentioned by Jex-Blake, Gard. E. Afr. ed. 4: 119 (1957).

Shrub 0.3–2.5 m tall, glabrous save for the young shoots. Leaves opposite, linear, 1.3–4.2(–6) cm long, 0.8–1.3(–2.5) mm wide, acute, with only the midrib evident, strongly revolute at margins, gland-dotted. Inflorescences 2–10-flowered, (0.3–)1–3 cm long, the flower pairs well-spaced. Petals pink, 5–6 mm diameter. Staminal claws purple, mauve, lilac, pink or white, 2.2–6.2 mm long with 30–90 filaments. Fruit globular, 7–11 mm wide with narrow opening.

8. **M. leucadendron** (*L.*) *L.* Of all the many sheets from East Africa identified as being this species only one appears to be correctly identified (Tanzania. Lushoto District: Amani, 18 Oct. 1928, *Greenway* 897).

Tree 15–40 m tall; young shoots silky with short adpressed hairs but soon glabrous. Leaves spirally arranged, narrowly lanceolate, 8.5–19.3 cm long, 0.9–2.5 cm wide, 9–14 times as long as wide, rather thinly coriaceous, acute or acutely acuminate, attenuate into the petiole, usually widest distinctly below the middle, soon glabrous. Spikes 1–3 together, terminal, also often solitary in upper axils forming a raceme-like collection of spikes, 6–15 cm long, 2.2–3 cm wide; rachis glabrous; flowers white or creamy white. Petals with small elliptic glands but usually no linear ones.

9. **M. quinquenervia** (*Cav.*) *S.T. Blake* (*M. leucadendron* auctt. *non* (L.) L. e.g. T.T.C.L.: 378 (1949), Jex-Blake, Gard. E. Afr. ed. 4: 118 (1957)). This has been widely cultivated in East Africa (Uganda. Mengo District: Entebbe Botanic Garden, Nov. 1930, *Snowden* 1815; Kenya. Nairobi, Karura Forest, *Battiscombe* 1006 (bis); Tanzania. Lushoto District: Amani Plantation 16, 26 June 1929, *Greenway* 1618 & Amani, New Nursery, 20 Mar. 1973, *Ruffo* 659 & Lushoto Arboretum, 26 June 1965, *Semsei* 3947; Kilosa District: Shamba Jovet, 11 Feb. 1926, *Burtt* 130).

Tree 6–25 m tall, 1.3 m circumference, with buff-coloured or white thickly laminated corky bark splitting into light papery flakes; young shoots densely silky. Leaves lanceolate to oblanceolate, 5–9 cm long, 0.6–2.4 cm wide, (3–)4–6(–7) times as long as wide, coriaceous, acute or narrowly obtuse, attenuate into the petiole, at length glabrous. Spikes solitary or 2–3 together, terminal and sometimes in uppermost 1–3 axils, 4–8.5 cm long, 2.5–3.5 cm wide; rachis sparsely to densely hairy or almost glabrous; flowers white, cream or pale yellow. Petals with linear and elliptic glands. Fig. 1/5–7, p. 4.

10. **M. bracteata** *F. Muell.* One specimen seen seems to be this species (Kenya, Nakuru District: Mugwathi Farm, Oct. 1967, *Shah in E.A.* 13856).

Shrub or small tree 2–15 m tall with dark grey fissured bark; young shoots pubescent. Leaves spirally arranged, sessile, very narrowly elliptic-lanceolate, 0.7–1.8(–2) cm long, 1–2.5 mm wide, up to 7-nerved, glabrous or pubescent. Inflorescence spiciform, terminal at first but axis soon grows out; flowers in 1–3-flowered groups, each in the axil of a persistent ovate bract, in the specimen seen, flowers appearing solitary and axillary. Calyx green, pubescent outside; petals and stamens slightly greenish cream. Style glabrous.

NOTE: The specimen was originally sent for naming since distillation had yielded an oil of possible economic value; it had been called *M. genistifolia* Sm. (correctly now *M. decora* (Salisb.) Britten), but it keys to *M. bracteata* in Byrnes's paper.

11. **M. styphelioides** *Sm.* has been grown in the Kenya Highlands. Nakuru District: Dundori Forest Station, 26 June 1959, *Pudden* 87; Kiambu District: Lari Arboretum, 12 Feb. 1953, *Darling* 61 & Limuru Arboretum, 11 Feb. 1972, *Kokwaro* 2994 & Closeburn, Oct. 1958, *Graham Bell in E.A.* 11531.

Shrub to 6 m or small tree unusually up to 20 m with typically layered papery bark. Leaves scattered, ovate-lanceolate, 0.4–2.5 cm long, 2–6 mm wide, acute with a pungent point, often twisted and concave above, 15–30-veined. Inflorescence few–many-flowered, dense, sometimes leafy; upper axillary, terminal or subterminal spikes with flowers single or in threes. Calyx-lobes with pungent points. Petals white, 1–2 mm long. Staminal bundles white with claw 3–4 mm long and 13–26 filaments.

12. **M. huegelii** *Endl.* (Kenya, Nairobi, 16 Apr. 1955, *Jex-Blake in E.A.* 10787).

Tree or shrub 1.2–6 m tall. Leaves small, ericoid, elliptic-lanceolate, 4–6 mm long, 1.7 mm wide, ± prominently 3–5-nerved. Stamens white. Fruiting spike narrow, 11.5 cm long, 8 mm wide, shorter in wild material.

13. **M. preissiana** *Schauer* is mentioned by Jex-Blake (Gard. E. Afr. ed. 4: 119 (1957)).

Shrub or tree 1.5–10 m tall with whitish papery bark. Leaves stiff, linear, 1.4 cm long, 1–2 mm wide, very narrowly rounded at the apex, very obscurely 1–3-nerved. Spikes terminal, 2.5–7 cm long, 2 cm wide; rachis silky pubescent; petals white or yellow, 2.5 mm long; staminal claws 2–3 mm long with (15–)16–30 filaments.

14. **M. lanceolata** *Otto* (*M. pubescens* Schauer) has been grown in Kenya. Nairobi, 16 Apr. 1955, *Jex-Blake in E.A.* 10787a.

Shrub or tree up to 8(–12) m tall, glabrous save for the young tomentose shoots; bark dark grey, fibrous. Leaves spirally arranged, very narrowly elliptic to ± lanceolate, 0.4–1.3 cm long, 1.1–1.5(–2.5) mm wide, mostly acute or mucronate, often rather thick, ± attenuate into a petiole 0.7–1.9 mm long; venation obscure. Inflorescences leafy spikes on glabrous or pubescent axes which grow on well before flowering, 2.2–4.8 cm long, of 6–14 groups of three flowers. Petals white, 2.5 mm long. Staminal claws white or cream, 0.5–1.8 mm long with 6–14 filaments.

15. **M. armillaris** (*Gaertn.*) *Sm.* This well known cultivated species has been grown in Kenya (Nairobi, Karen, Hort. Gardner, 25 July 1973, *Gardner in E.A.* 15385) and Tanzania (Arusha District: Oldonyo Sambu, Schutz Farm, 10 Oct. 1945, *Van Rensburg* 157) (Jex-Blake, Gard. E. Afr. ed. 4: 119 (1957)).

Graceful shrub or small tree 0.5–9 m tall with furrowed grey bark which peels from lower part of trunk; branches pendulous; characteristic white lines on shoots from which leaves originate. Leaves spirally arranged, needle-like, 0.9–2.8(–3.5) cm long. Flowers white in cylindrical spikes 2.5–6.5 cm long, 2.5 cm wide. Staminal claw 4–8 mm long with 10–20 white or rarely mauve filaments.

16. **M. nesophila** *F. Muell.* has been grown in Kenya. Kiambu District: Closeburn, July 1956, *Graham Bell & Greenway* 15.
Shrub 1–7 m tall, ± glabrous. Leaves alternate, obovate-oblanceolate to oblong-elliptic, 1–2(–4) cm long, 3–9 mm wide, rounded and sometimes mucronulate at the apex, conspicuously wrinkled but nervation often scarcely visible. Flowers in terminal heads 1.8–2.5 cm diameter, usually overtopped by outgrowing leafy axes after anthesis. Petals ± 2 mm long, scarious. Staminal claws about same length with 10–15 pink, lavender or purple-red filaments.
NOTE: Some of the workers on this genus in Australia have used very different concepts of leaf-shapes, calling some narrowly elliptic or ovate which to me are far too narrow to merit these terms.

ANGOPHORA *Cav.*
Angophora costata (*Gaertn.*) *Britten* (*A. lanceolata* (Pers.) Cav.) has been grown in Kenya (Nairobi Arboretum, 4 Dec. 1964, *Noronha in E.A.* 13058, also at Kaptagat and Londiani).
Tree to 30 m with pink or pinkish grey smooth bark. Leaves very like those of some *Eucalyptus* species, narrowly lanceolate or elliptic to narrowly ovate, (4–)8–19 cm long, (0.6–)1.2–3.5 cm wide, acute at the apex, attenuate at the base, glabrous. Flowers in 3-flowered umbels with axes covered with stiff capitate hairs or ± glabrous. Calyx lobes 5, up to 3 mm long, similarly hairy. Petals white, ± round, ± 4 mm diameter. Fruits campanulate or ovoid, woody, 1–1.7 cm long and wide, prominently ribbed, glabrous to densely hairy.

I have a note that *A. subvelutina* F. Muell. has also been cultivated in Nairobi.

LOPHOSTEMON *Schott.*
Lophostemon confertus (*R. Br.*) *P.G. Wilson & J.T. Waterhouse* (*Tristania conferta* R. Br., by which name it is much better known) has been widely grown in East Africa and is often called Brisbane box (Uganda. Mengo District: Entebbe Botanic Gardens, fide Dale. Kenya. Nairobi, Arboretum, 15 Dec. 1952, *Greenway* 8745; Kiambu District: Muguga?, *Pudden* 11; Central Kavirondo District: Maseno, 3 Oct. 1952, *Abraham* 9; Tanzania. Bukoba District: Dec. 1967, *Kanywa* 31; Lushoto District: Amani, Drackensberg Plantation 6, 17 July 1930, *Greenway* 2258 & Lushoto Arboretum, 12 Aug. 1970, *Ngoundai* 438; Njombe District: Forest Dept. trial plot 2, Jan. 1962, *Procter* 1988, also Nairobi, Nyeri and Kaptagat) (T.T.C.L.: 380 (1949); Dale, Introd. Trees Uganda: 69 (1953); Jex-Blake, Gard. E. Afr. ed. 4: 244 (1957)).
Tree 5–30 m tall, 1.3 m girth, with longitudinally furrowed stringy bark. Leaves in whorls of 4–5, elliptic to lanceolate, 7–17 cm long, 2–6.5 cm wide, acute to acuminate at apex. Flowers scented, in few-flowered axillary cymes. Calyx long white-hairy; lobes narrow and acute. Petals white. Stamens in 5 bundles with linear claws as long as petals. Fruit hemispherical to cup-shaped, 7–10 mm long, 9–12 mm wide. Fig. 3, p. 12.

L. suaveolens (*Gaertn.*) *P.G. Wilson & J.T. Waterhouse* (*Tristania suaveolens* (Gaertn.) Sm.) (T.T.C.L.: 380 (1949); Dale, Introd. Trees Uganda: 69 (1953)) is much less frequently grown (Uganda fide Dale; Tanzania. Lushoto District: Amani, *Greenway* 2863).
Tree to 11 m differing in short obtuse calyx-lobes and staminal claws being half as long as the petals.

SYNCARPIA *Ten.*
Syncarpia glomulifera (*Sm.*) *Nied.* (*S. laurifolia* Ten.) is widely grown in East Africa (Uganda. Ankole District: Mbarara, Dec. 1934, *Harris* 198; Ankole, Gayaza, Aug. 1936, *Eggeling* 3231. Kenya. ?Muguga, *Pudden* 10. Tanzania. Lushoto District: Amani, Plantation 18, 24 June 1929, *Greenway* 1598 & Lushoto, Aug. 1955, *Semsei* 2342; Morogoro District: Bunduki, 3 Jan. 1948, *Wigg* 2282).
Tree 3–20 m tall with buttressed trunk and very rough reddish peeling stringy bark; youngest parts with long white hairs. Leaves narrowly oblong-elliptic to elliptic, 4.5–13 cm long, 1.7–5.5 cm wide, narrowing to a narrowly rounded apex, cuneate to rounded at the base, the apical ones often appearing in pseudowhorls of 4 or 6, finely tomentose beneath, very closely

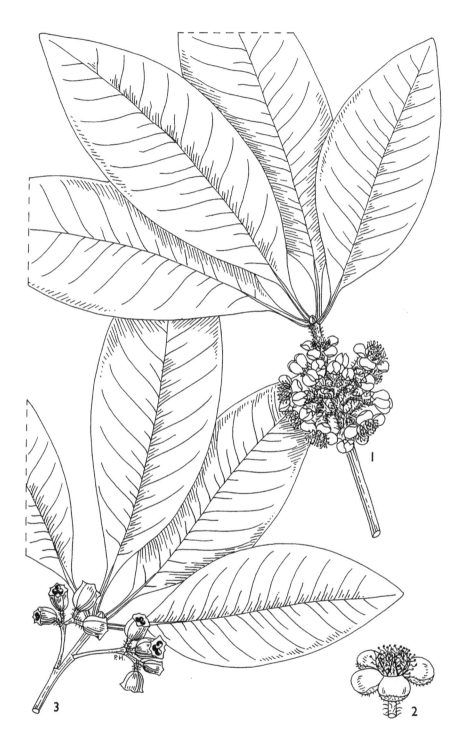

FIG. 3. *LOPHOSTEMON CONFERTUS*— **1**. habit, × ²/₃; **2**, flower, × 1.3; **3**, habit in fruit, × ²/₃. 1–2
from *Greenway* 8745; 3 from *Ngoundai* 438. Drawn by Pat Halliday.

reticulate and ± glabrous above. Inflorescences fused globose heads 1.5–2 cm diameter, the calyx-tubes fused, grey-tomentose; peduncles 2–5 cm long. Fig. 4/6, p. 14.

NOTE: The name *Syncarpia procera* (Salisb.) Domin in Biblioth. Bot. 22: 1026 (1928), based on *Metrosideros procera* Salisb., Prodr. Stirp. Hort. Chapel Allerton: 351 (1796), type: Australia, Port Jackson, *D. Burton* s.n., has been used for this species by a few Australian authors. It predates the basionym of Niedenzu's name by a few months: *Metrosideros glomulifera* Sm. in Trans. Linn. Soc., London 3: 269 (1797), type: Australia, Port Jackson, *D. Burton* s.n. (BM-herb. Banks!, holo., LINN-herb. Smith 877.11, iso., microfiche!). I have not seen Salisbury's type and I am content to follow most recent Australian authors and use *S. glomulifera*. The problem was discussed by J. Britten as long ago as 1916, in J.B. 54: 62. Now that specific names can be conserved it would be possible to put up *S. glomulifera* for conservation, if this was found necessary.

METROSIDEROS *Gaertn.*

Metrosideros* excelsa *Gaertn.* (*M. tomentosa* A. Rich.). A native of New Zealand where it is often called the Christmas-tree, it is recorded as cultivated in Kenya at Elburgon (source of record lost) but no material has been seen.

Tree to 21 m with short thick trunk to 1.5 m diameter. Leaves lanceolate to oblong, 2.5–10 cm long, 1–5.2 cm wide, obtuse or acute at the apex, thick, usually tomentose beneath. Flowers dark crimson in many-flowered terminal cymes with stamens 3–3.8 cm long and calyx and pedicels densely grey tomentose.

ACCA *O. Berg*

Acca sellowiana (*O. Berg*) *Burret* (much better known as *Feijoa sellowiana* (O. Berg) O. Berg) feijoa or pineapple guava, a native of South America, is now widely grown in the tropics and warm climates (particularly New Zealand) as an ornamental, and for its fruit (Dale, Introd. Trees Uganda: 41 (1953); Jex-Blake, Gard. E. Afr. ed. 4: 113, 299, 345 (1957); F. White in F.Z. 4: 185 (1978); Morton, Fruits Warm Climates: 367–370, fig. 100 (1987)). (Kenya. Nairobi Arboretum, 12 Feb. 1952, *Williams Sangai* 332 & Nairobi, July 1953, *Bally* 9049 & 29 Oct. 1962, *Lucas* H293/62 & Karen, Hort. Gardner, 29 July 1973, *Gillett* 20298; Tanzania. Lushoto District: Amani, 31 Oct. 1969, *Ngoundai* 432).

Shrub or small tree 0.9–7 m tall with pale grey bark. Leaves opposite, elliptic, or oblong-elliptic, 2.8–7 cm long, 1.5–4 cm wide, obtuse at the apex, smooth and glossy olive-green above, densely white-tomentose beneath. Flowers in 1–3-flowered axillary or terminal cymes. Calyx-tube and outside of lobes white-tomentose. Petals white or pink outside, crimson-purple or lilac-pink inside, up to 2 cm long. Stamens 1.5–2.5 cm long, with crimson or purple filaments and yellow anthers. Fruit dull green or yellow-green, often with dull red or orange blush, oblong or ellipsoid-ovoid, 2.5–8 cm long, 2.8–5 cm wide. Fig. 4/4–5, p. 14.

NOTE: Occasionally abnormal flowers occur with 7 calyx-lobes and petals instead of 4, and 5 fleshy bodies instead of 1 style.

MYRTUS *L.*

Myrtus communis *L.* The well known myrtle of the Mediterranean, widely cultivated throughout the world for ornament, grows well in East Africa (F.F.N.R.: 303 (1962); T.T.C.L.: 378 (1949), var. *tarentina* L.); Jex-Blake, Gard. E. Afr. ed. 4: 120 (1957); Friis in Fl. Eth. 2 (2): 71, fig. 72/1 (1996)) (Tanzania. Lushoto District: Amani, *Greenway* 2307).

Shrub 0.9–3 m tall or sometimes a small tree, with scented foliage. Leaves lanceolate, oblong-lanceolate or ovate, 0.5–2 cm long. Flowers white, solitary, axillary. Berry blue-black. Fig. 4/1–3, p. 14.

PIMENTA *Lindl.*

Pimenta dioica (*L.*) *Merr.* (*Myrtus pimenta* L., *M. dioica* L., *Pimenta officinalis* Lindl.) (T.T.C.L.: 378 (1949); U.O.P.Z.: 412 (1949); Dale, Introd. Trees Uganda: 55 (1953)). Allspice or pimento, a native of Jamaica, has been grown at several places in East Africa (Uganda. Mengo District: Entebbe, fide Dale; Tanzania. Lushoto District: Amani, 21 Nov. 1921, *Soleman A.H.* 6053 & Plantation 16, 26 June 1929, *Greenway* 1625; recorded from Zanzibar).

---

* Recent attempts to treat this name as masculine are in error. Gaertner originally had it feminine and there seems no reason to change it.

FIG. 4. *MYRTUS COMMUNIS* — **1**. habit in fruit, × ²/₃; **2**, habit, × ²/₃; **3**, fruit, × 1. *ACCA SELLOWIANA* — **4**, habit, × ²/₃; **5**, fruit, reduced. *SYNCARPIA GLOMULIFERA* — **6**, habit, × ²/₃. 1–3 from *Col. R. Price*, cult. Richmond, Surrey, U.K. & cult. Hort. Kew (on 1 sheet); 4–5 from live material cult. by *R.D. Meikle*, Somerset, U.K. and from *Gillett* 20298; 6 from *Harris* 198. Drawn by Pat Halliday.

Tree to 20 m with young branchlets glandular and pubescent. Leaves aromatic, oblong-elliptic, elliptic or elliptic-lanceolate, 5.5–17(–22) cm long, 2–6.5(–8) cm wide, rounded to obtusely acuminate at the apex, ± glabrous beneath. Flowers white, 4-merous, small in many-flowered panicles 5–12 cm long. Fruit subglobose, 5–10 mm diameter, densely covered with convex glands. Fig. 1/3–4, p. 4.

P. racemosa (*Mill.*) *J.W. Moore* (U.O.P.Z.: 412 (1949)) (*Amomis caryophyllata* (Jacq.) Krug & Urb. (T.T.C.L.: 370 (1949); Dale, Introd. Trees Uganda: 6 (1953)); *Pimenta acris* (Sw.) Kostel.). Bayrum tree, native of West Indies and northern South America, has been grown in Tanzania (Lushoto District: Sigi, 15 Feb. 1921, *Soleman A.H.* 6054 & Sigi Chini 16, 9 Dec. 1940, *Greenway* 6071 & Amani Chini Plantation, 23 Mar. 1973, *Ruffo* 657 & 1051; has been recorded from Zanzibar and Pemba).

Shrub or small slender tree 7.5–15 m tall; bark thin and scaly, peeling to produce a mottled trunk; essentially glabrous but with dense glands on stems and leaves. Leaves very aromatic, obovate to elliptic, 3–15 cm long, 1.2–7.5 cm wide, acute to broadly rounded at the apex, coriaceous. Flowers white, 5-merous, small, in many flowered ((5–)15–100) panicles 2.5–12 cm long. Fruit subglobose, 6–12 mm diameter, 1–4-seeded (2–8 fide T.T.C.L.).

NOTE: The nomenclatural complications of the last two species are explained by Landrum, Fl. Neotropica, Mon. 45: 79, 83, 107 (1986). He also figures both species, fig. 24 & 45.

MYRCIARIA *O. Berg*
M. floribunda (*Willd.*) *O. Berg* (*Eugenia floribunda* Willd., *E. protracta* Steud.) the rumberry or guava berry has been grown for its edible fruits. Sobral in Napaea 9: 16 (1993) details the amazingly voluminous synonymy of this species and Morton, Fruits of warm climates: 388 (1987) describes it. Tanzania. Lushoto District: Amani, 12 July 1946, *Greenway* 7846 & Kiumba, 28 May 1946, *Don Carlos* in *A.H.* 9822 & Amani, Drackenberg Plantation, 12 July 1946, *Greenway* 7845 & Amani, *Furuya* 254. A native of Cuba, Hispaniola, Jamaica, Porto Rico etc. it also occurs in S Mexico, Belize, Guatemala, Salvador, the Guianas, E Brazil and N Colombia.

Evergreen shrub or small tree 3–4.5 m tall with reddish brown branchlets, puberulous when young; bark flaking. Leaves elliptic-lanceolate, 2.5–8 cm long, 0.8–3 cm wide with slender attenuate apex with sharply acute tip; petiole 3 mm long. Flower clusters subsessile. Stamens ± 75. Fruit orange to dark red or blackish red, globose or ± oblate-spheroid, 0.8–1.6 cm diameter.

EUCALYPTUS *L'Hér.*, Sert. Angl.: 18 (1788); Chippendale, Fl. Austral. 19: 1–448 (1988)

Trees or shrubs with smooth to very rough bark of various types. Leaves usually markedly different in juvenile, intermediate and adult phases, the juveniles often opposite, cordate and sessile but adults mostly alternate, usually petiolate, falcate and pendulous. Flower usually in umbel-like condensed usually pedunculate dichasial cymes, either single or more rarely paired in the leaf-axils or in terminal corymbose panicles; flowers 3 or more per umbel or in a few species solitary, sessile or pedicellate. Calyx-limb and corolla each or together forming an operculum which is shed at anthesis but calyx-lobes sometimes free, falling separately or together. Stamens numerous; anthers dehiscing by pores or slits; connective mostly with a dorsal or terminal gland. Ovary 2–7-locular, inferior or partly superior; ovules numerous. Capsule usually woody, loculicidal or rarely circumscissile with scars left by falling operculum and androecium; disc convex or descending; valves included to exserted.

Over 500 (possibly 900) species with new ones constantly being described, restricted to Australia save for some 10 species in Malesia to the east of Wallace's line. The species hybridise easily (within the subgenera) and this adds to the already considerable taxonomic difficulty of this genus. Over a third of the species have been cultivated outside their natural range both in the tropics and temperate regions. Hybrids arise between species not growing together naturally and which are therefore unknown in Australia. Eucalypts appeal to foresters since not only do most grow quickly, but there are species adapted to a very wide range of environments. In the early 1960's extensive trials were carried out in Uganda (Ann. Rep. For. Dept. 1963/4: 15–16 (1965)) using different species of *Eucalyptus* – *E. saligna, E. grandis* and *E. deglupta* in grasslands associated with moist semi-deciduous forest around Lake Victoria; *E.*

saligna, E. microcorys and *E. paniculata* in the western *Pennisetum* grasslands; *E. camaldulensis* and
E. maidenii in woodland and wooded grassland at under 1500 m; *E. saligna* in wooded grassland
over 1500 m; *E. globulus* and *E. microcorys* in montane scrub; *E. saligna, E. grandis* and *E. deglupta*
in moist semi-deciduous forest up to 1500 m; *E. globulus* in similar forest and montane forest
at higher altitudes.

Friis (Fl. Eth. 2 (2): 71–106 (1996) ("1995")) deals fully with all the species recorded from
Ethiopia. Mullin in Kirkia 16: 95–107 (1997) ("1996") gives a list of 199 taxa which have been
introduced into Zimbabwe (no hybrids are mentioned). *Corymbia* is mentioned as a subgenus,
but this was split from *Eucalyptus* and described as a genus of 113 (many new) species, by Hill &
Johnson in 1995 (Telopea **6**: 185–504 (1995)). I can find no valid use of it as a subgenus but it
is mentioned by Pryor & Johnson, A Classification of the Eucalypts: 18, 28, 34, 35 (1971). Their
new classification was deliberately divorced from the Code of Nomenclature so the name is not
a basionym of Hill & Johnson's genus although a development of previous work. For a local
flora *Eucalyptus* has been maintained without implying any taxonomic decision. Alternative
names in *Corymbia* are indicated.

It is difficult to know how to treat the genus here – at least 130 species have been cultivated
in East Africa but many are only to be found in Forestry Dept. plantations and experimental
plots but about 50 are more widespread and some very widely cultivated and form a
conspicuous part of the landscape. Many are used as very rapidly growing trees for timber and
firewood, also for shade and ornament. These more widespread species have been keyed and
briefly described (marked with ▼ in list). A list is also given of all the species of which I have
seen authentic material. In 1963 I produced a cyclostyled edition of a key* I had made based
on the material in the East African Herbarium. It was not really meant to enable every
Eucalyptus to be named in the field but to cut down the time taken to name them by comparison
in the herbarium. The herbarium material had been named as far as possible by staff at the
Forestry and Timber Bureau in Canberra but of course many changes have been made since
then. I am very grateful to Ken Hill, who was fortunately Australian Liaison Officer at Kew when
I was writing this account, for naming many specimens for me; also to Ian Brooker for reading
through the account and correcting many errors and making additions. He also suggested that
parts of the key where I had optimistically attempted to key out hybrids should be omitted.

List of *Eucalyptus* known to have been cultivated in East Africa** (usually only listed if authentic
specimens seen).

            E. acmenoides Schauer; Kenya: Nairobi, Muguga (named *triantha* Link but that is of
               dubious identity)
            E. alba (Reinw.) Blume; Tanzania: Lushoto
▼   *E. albens* Benth.
            E. amygdalina Labill., 1806 (?*E. salicifolia* Cav., 1797)***
            E. astringens Maiden; Kenya: Nairobi, Lari; Tanzania: Lushoto
            E. baxteri (Benth.) J.M. Black; Kenya: Muguga
▼   *E. bicostata* Maiden, Blakely & J.H. Simmonds
▼   *E. blakelyi* Maiden
▼   *E. bosistoana* F. Muell.
▼   *E. botryoides* Sm.
            E. botryoides Sm. × *E. grandis* Maiden; Kenya: Muguga, Mwanda
            E. bridgesiana R.T. Baker; Kenya: Muguga
            E. brockwayi C.A. Gardner; Tanzania: Lushoto
            E. caesia Benth.; Nairobi: Karen
▼/©   *E. calophylla* Lindl.
▼   *E. camaldulensis* Dehnh. (the most widely cultivated species)

* A key to the species of *Eucalyptus* held in the East African Herbarium, Nairobi. Nairobi,
1963 (cyclostyled).
** Those dealt with more fully are marked with an ▼; © indicates the taxon belongs to the
genus *Corymbia* K.D. Hill & L.A.S. Johnson
*** If it can be proved that these two are synonymous then *E. amygdalina* should be put up
for conservation since it is used in recent standard works (see also Blake in Austral. Journ.
Bot. 1: 306–7 (1953))

        E. camaldulensis Dehnh. × *E. grandis/saligna*

        E. camphora R.T. Baker; Kenya: probably Muguga

        E. canaliculata Maiden; Kenya: Lumbwa, Segeria

        E. capitellata Sm.; Kenya: Molo (aff.)

        E. cephalocarpa Blakely; Tanzania: Amani

        E. cinerea F. Muell.; Kenya: Nyahururu [Thomson's Falls]

        E. cinerea F. Muell. × *E. bridgesiana* R.T. Baker; Tanzania: Amani (probably)

▼/©   *E. citriodora* Hook.

▼   *E. cladocalyx* F. Muell.

        E. cloeziana F. Muell; Kenya: Muguga (not authenticated)

        E. cornuta Labill.; Kenya: Nairobi

▼   *E. crebra* F. Muell.

        E. crucis Maiden; Kenya: Nairobi

        E. cypellocarpa L.A.S. Johnson; Kenya: Muguga

        E. dalrympleana Maiden; Tanzania: Amani

        E. deglupta Blume; Kenya: Muguga; Tanzania: Rau

        E. delegatensis R.T. Baker subsp. *tasmaniensis* Boland; Kenya: Muguga, Lari (not authenticated) (= *E. gigantea* Hook.f.)

▼   *E. diversicolor* F. Muell.

        E. dives Schauer; Kenya: Muguga, Lari

▼   *E. drepanophylla* F. Muell.

        E. elata Dehnh.; Tanzania: Amani (*E. andreana* Naudin)

        E. erythrocorys F. Muell.; Kenya: Nairobi, Muguga

        E. erythronema Turcz.; Kenya: Muguga (not authenticated)

        E. eugenioides Spreng.; Kenya: Nairobi (= *E. acervula* Sieber)

©   *E. eximia* Schauer; Tanzania: Amani

        E. falcata Turcz.; Kenya: Muguga (not authenticated)

▼   *E. fastigata* H. Deane & Maiden

        E. fibrosa F. Muell.; Kenya: Ngong

        E. fibrosa F. Muell. subsp. *nubila* (Maiden & Blakely) L.A.S. Johnson; Kenya: Muguga, Ngong (*E. nubila* Maiden & Blakely, "*nubilis*")

▼/©   *E. ficifolia* F. Muell.

        E. flocktoniae (Maiden) Maiden; Kenya: Muguga (not authenticated)

        E. gardneri Maiden; Kenya: Muguga, Lari (not authenticated)

        E. gillii Maiden; Kenya: Muguga (not authenticated)

▼   *E. globulus* Labill.

▼   *E. gomphocephala* DC.

▼   *E. grandis* Maiden

        E. grandis Maiden × *E. botryoides* Sm.

        E. grandis Maiden × *E. tereticornis* Sm.; Tanzania: Lushoto

        E. grossa Benth.; Kenya: Nairobi

▼/©   *E. gummifera* (Gaertn.) Hochr.

        E. gunnii Hook.f.; Tanzania: Amani ('probably')

©   *E. intermedia* R.T. Baker; Kenya: Muguga (not authenticated)

        E. cf. *johnstonii* Maiden; Tanzania: Lushoto

        E. largiflorens F. Muell.; Kenya: Machakos, Nairobi, Muguga (*E. bicolor* Mitch.)

        E. latifolia F. Muell.; Tanzania: Rau Forest

        E. lehmannii (Schauer) Benth.; Kenya: Nairobi, Muguga; Tanzania, Amani

        E. lesouefii Maiden; Tanzania: Lushoto

▼   *E. leucoxylon* F. Muell.

        E. leucoxylon F. Muell. × *E. sideroxylon* Woolls

        E. longicornis (F. Muell.) Maiden × *E. tereticornis*. Kenya: Lari (dubious)

▼   *E. longifolia* Link

        E. loxophleba Benth.; Kenya: Nairobi

        E. macarthurii H. Deane & Maiden; Kenya: Muguga

        E. macrocarpa Hook.; Kenya: Karen

        E. macrorrhyncha F. Muell.; Kenya: Muguga, ? Eldoret; Tanzania: ? Amani

▼/©   *E. maculata* Hook. (very widely cultivated)

▼   *E. maidenii* F. Muell.

        E. mannifera Mudie; Tanzania: Lushoto

        E. marginata Sm.; Kenya: Muguga; Tanzania: ? Amani

        E. megacarpa F. Muell.; Kenya: Nairobi

        E. melanophloia F. Muell.; ? Kenya: Nairobi, Lari

E. cf. *melanoleuca* S.T. Blake; Kenya: no data
▼   E. *melliodora* Schauer
▼   E. *microcorys* F. Muell.
▼   E. *microtheca* F. Muell.
E. *moluccana* Roxb.; Tanzania: Amani ('probably') (*E. hemiphloia* F. Muell.)
▼   E. *muelleriana* Howitt
E. *nicholii* Maiden & Blakely; Kenya: no data
E. *notabilis* Maiden; Kenya: Kikuyu
E. *nova-anglica* H. Deane & Maiden; Kenya: Muguga
▼   E. *obliqua* L'Hér.
E. *occidentalis* Endl. × *E. gomphocephala* DC; Kenya: Nairobi
E. *oleosa* Miq.; Kenya: Nairobi; Tanzania: Lushoto (? var. *obtusa* C.A. Gardner)
E. *ovata* Labill.; Kenya: Muguga; Tanzania: Dar es Salaam
E. *oxymitra* Blakely; Tanzania: Lushoto
E. *pachyphylla* F. Muell.; Tanzania: Lushoto
▼   E. *paniculata* Sm. (*E. fergusonii* R.T. Baker) (very widely cultivated)
E. *patentinervis* R.T. Baker; Kenya: Londiani, Kisii ("aff.") (hybrid between *E. robusta* and
      E. tereticornis)
E. *pauciflora* Spreng.; Kenya: Lari; Tanzania: Lushoto
▼   E. *pellita* F. Muell.
E. *pilligaensis* Maiden; Kenya: Machakos
▼   E. *pilularis* Sm.
E. *piperita* Sm.; Tanzania: Lushoto, Kwai
E. *planchoniana* F. Muell.; Uganda: Hoima
E. *platyphylla* F. Muell.; Tanzania: Lushoto, Dar es Salaam, ?Dodoma (Kigwe)
▼   E. *polyanthemos* Schauer
E. *populnea* F. Muell.; Tanzania: Maswa, Lushoto
E. *preissiana* Schauer; Kenya: Nairobi
E. *propinqua* H. Deane & Maiden; Kenya: Nyeri, Bahati, Kaptagat; Tanzania: ? Amani
E. *pulchella* Desf.; Kenya: Nairobi, Londiani (*E. linearis* Dehnh.)
▼   E. *punctata* DC.
E. *radiata* Sieber; Tanzania: ? Amani
E. *radiata* Sieber subsp. *robertsonii* (Blakely) L.A.S. Johnson & Blaxell; Kenya: Muguga
E. *redunca* Schauer var. *elata* Benth.; Kenya: Lari (not authenticated); Tanzania: Lushoto
▼   E. *regnans* F. Muell.
▼   E. *resinifera* Sm.
▼   E. *robusta* Sm. (very widely cultivated)
E. *robusta* × *E. saligna*; Kenya: Macalder Mines
E. *robusta* ? × *E. botryoides*; Kenya: Londiani
E. *robusta* ? × *E. diversicolor*; Kenya: Nanyuki
E. *rossii* R.T. Baker & H.G. Sm.; Kenya: Nairobi
E. *rubida* H. Deane & Maiden; Kenya: Muguga (not authenticated)
▼   E. *rudis* Endl.
E. *rudis* ? × *E. camaldulensis*; Kenya: Rumuruti
▼   E. *saligna* Sm.
E. *saligna* × *E. grandis*; Kenya: Maseno
E. *salmonophloia* F. Muell.; Kenya: Nairobi, Muguga; Tanzania: Lushoto
E. *salubris* F. Muell.; Kenya: Nairobi, Muguga, Lari
E. *scabra* Dum.Cours.; Kenya: Muguga (not authenticated); Tanzania: Amani (Fl.
      Austral. says identity doubtful)
E. *seeana* Maiden; Kenya: Muguga (not authenticated)
▼   E. *siderophloia* Benth.
▼   E. *sideroxylon* Woolls
E. *sieberi* L.A.S. Johnson; Kenya: Muguga
E. *socialis* F. Muell.; Tanzania: Lushoto
E. *spathulata* Hook.; Kenya: Muguga (not authenticated)
E. *staigeriana* Bailey; Kenya: Nairobi, Lari
E. *staigeriana* × *E. crebra*; Kenya: Nairobi, Lari
E. *stricklandii* Maiden; Kenya: Nairobi, Muguga
▼   E. *tereticornis* Sm. (very commonly cultivated)
E. *tetragona* (R. Br.) F. Muell.
E. *torquata* Luehm.; Kenya: Nairobi, Muguga; Tanzania: Lushoto
©   E. *torelliana* F. Muell.; Uganda: Entebbe; Kenya: Mombasa; Tanzania: Ukerewe, Lushoto

E. × *trabutii* H. Vilm.; Kenya: Nairobi (reputedly cultivated hybrid between *E. botryoides* and *E. camaldulensis*)

E. transcontinentalis Maiden; Kenya: Muguga

▼   *E. viminalis* Labill.

E. cf. *viridis* R.T. Baker

E. woollsiana R.T. Baker; Kenya: Muguga (*E. microcarpa* (Maiden) Maiden)

E. sp.: *Braun* in *A.H.* 3358 (Amani, 11 Feb. 1911) has not been recognised by any Australian workers. Flowers resemble northern forms of *E. camaldulensis* but leaves are quite different.

?*E. sp.*: *Magogo* 181 (Lushoto District: Htazina, 15 Nov. 1971) - mature flowers and fruit are needed. Ken Hill has suggested this is not a *Eucalyptus* because of the staminal glands with hanging anthers.

KEY TO THE MOST COMMONLY CULTIVATED SPECIES OF *EUCALYPTUS*

This key is based mainly on cultivated East African material and only on Australian material where certain stages of the former were not available. The measurements, particularly the ranges, differ from those of the descriptions that follow (which are based on both cultivated material and the Flora of Australia).

This key is no more than a guide to the species commonly planted. There is quite a high chance that the plant being keyed is not in the key. For this reason the information given in the couplets is not always contrasting; items are often added to help eliminate doubt. Recourse must be had to Australian literature for less frequently introduced species but this may not work on the odd hybrids grown in East Africa or variants due to quite different growth conditions in East Africa.

1. Flowers solitary (rarely in threes) in the leaf-axils — 15. *E. globulus*
   Flowers in threes or > 3, in umbels or panicles of umbels · · · · 2
2. Mature fruits exceeding 20 mm long and/or wide · · · · 3
   Mature fruits not exceeding 20 mm · · · · 4
3. Usually a small ornamental tree up to 10 m with red, pink or orange flowers; seed winged · · · 14. *E. ficifolia*
   Tree attaining 40(–60) m with cream (rarely pink) flowers; seed not winged; not used as an ornamental · · · · · · 6. *E. calophylla*
4. Fruits with one dimension about 13–16(–18) mm · · · · 5
   Fruits not exceeding 13(–13.5) mm in any direction · · · · 11
   (some species have been included in the key twice)
5. Flowers in apparently lateral panicles; pedicels 10 mm long; peduncles 20 mm long; fruits 8–13.5 × 11.5 mm · · · · 14
   Flowers in simple umbels or in apparently terminal panicles · · · · 6
6. Peduncles ± round; fruits distinctly pendulous, campanulate, 10–17 × 9–16 mm; pedicels and peduncles 13–20 mm long · · · · · 20. *E. longifolia*
   Peduncles angled or flattened · · · · 7
7. Valves included (sometimes very slightly exserted and joined across orifice in *E. robusta* particularly in not quite mature fruits) · · · · 8
   Valves flush to strongly exserted · · · · 9

8. Operculum hemispherical-conic, ± obtuse, not
   long-drawn-out; fruits elongate-urceolate,
   13–15(–20) × 7–10.5(–15) mm (high values
   from Fl. Austral.); pedicels 3–12(–15) mm
   long; peduncles 25–35 mm long · · · · · · · ·1   8. *E. gummifera*
   Operculum elongate-conical, rostrate, often
   quite long-drawn-out; fruits narrowly ovoid-
   cylindric or cylindric-urceolate, (8.5–)9–15
   (–18) mm long, 6–12 mm wide; pedicels 1–12
   mm long; peduncles 13–30(–35) mm long · · ·   35. *E. robusta*

9. Stems quadrangular; peduncles 0–7 mm long,
   angular; fruits bicostate, 14–17 × 14–20 mm
   with disc broad smooth and thick, sometimes
   concealing the short thick valves · · · · · · · · ·   3. *E. bicostata*
   Stems not quadrangular and without other
   characters combined ·10

10. Fruits campanulate, 8–16 × 10–15 mm with a
    very characteristic inset oblique disc 2.5–4
    mm high (fig. 9/29); valves strongly exserted
    (or rarely just included); pedicels thick,
    2(–12) mm long; peduncles ± 12 mm long,
    less flattened, ± 5 mm wide · · · · · · · · · · · · ·   29. *E. pellita*
    Fruits oblong-turbinate, 16–18 × 12–15 mm
    with low convex disc fused to flush or less
    strongly exserted valves; peduncles 18–30
    mm long, very flat, 7.5–8 mm wide; buds
    characteristic (fig. 7/16) · · · · · · · · · · · · ·   16. *E. gomphocephala*

11. Valves included, often deeply ·12
    Valves exserted or at least flush (this character
    does vary e.g. in *E. pellita* some fruits can have
    valves just included) · 34 (p. 23)

12. Flowers mostly in terminal or lateral panicles · · · · · · · · · · · · · · · · · · ·13
    Flowers mostly in axillary umbels – can appear
    paniculate but overtopped by leafy branch* · · · · · · · · · · · · · · · · · · ·21

13. Fruits over 10 mm long, urceolate ·14
    Fruits under 10 mm long or, if ± 10 mm, then · · · · · · · · · · · · · · · · · · ·
    not urceolate or only slightly so ·15

14. Fruits 10–12(–15) × 9–11 mm; foliage strongly
    lemon-scented, usually narrow-leaved · · · · ·   8. *E. citriodora*
    Fruits 8–13.5 × 8.5–11.5 mm; foliage not lemon-
    scented, usually broader-leaved, otherwise
    almost identical with last · · · · · · · · · · · · ·   21. *E. maculata*

15. Mature leaves ovate to lanceolate, 5.5–9 ×
    1.5–3.5 cm (juvenile round); fruits turbinate,
    4.5–5 × 4–4.5 mm; peduncles 10 mm long;
    pedicels ± 7 mm long · · · · · · · · · · · · · · · · ·   31. *E. polyanthemos*
    Mature leaves mostly lanceolate (but immature
    can be ovate) ·16

16. Fruits mostly larger, 6 mm or more long ·17
    Fruits mostly under 5 mm long or scarcely
    exceeding 5 mm ·19

* See under *E. cladocalyx* p. 33; other species may be similar e.g. *E. diversicolor.*

17. Branches, buds and fruits etc. very glaucous; fruits cylindric-urceolate or barrel-shaped, 7–10 × 6–8 mm · · · · · · · · · · · · · · · · · · ·   1. *E. albens*

     Branches etc. not glaucous; fruits not as above · · · · · · · · · · · · · · · · · · · 18

18. Flowers with a shallow calyx which tapers characteristically into the pedicel; fruits oblong or turbinate-campanulate, 6–9 × 4.5–5.5 mm; pedicels 9–10 mm long; peduncles mostly round; leaf-margins subundulate and leaves discolorous; venation rather open with obvious marginal nerve; bark fibrous · · · · · · · · · · ·   24. *E. microcorys*

     Flowers with a deeper calyx; fruits ovoid-turbinate to hemispherical, 5–8.5 × 5.5–8 mm, sometimes 2-ribbed; pedicels 2–10 mm long, thickened above, tapering below or ± winged; peduncles flattened, subangular or ± rounded; leaf-margins straighter but leaves often discolorous; ironbark · · · · · · · · · · · ·   38. *E. paniculata*

     (some specimens of *E. drepanophylla* could key here)

19. Box-like bark with rough finely fibrous bark on lower trunk or ± smooth throughout; juvenile leaves broadly lanceolate to ovate or round; fruit campanulate 3.5–5 × 5 mm; pedicels 3(–8) mm long; peduncles 6(–10) mm long; valves 5 or 6 ·   4. *E. bosistoana*

     Ironbark with rough dark grey to black bark throughout or fibrous grey-yellow or red-brown bark persistent to variable heights and smooth and pale yellowish above · 20

20. Flowers larger, 8 × 5–6 mm; fruits campanulate, slightly narrowed at apex, 5–5.5 × 4.5–6 mm; pedicels 5–8 mm long; peduncles 15 mm long; fruit surface reticulate; venation ± open; valves 4 or 5; bark fibrous (as in couplet 19/2) ·   23. *E. melliodora*

     Flowers smaller, 4.5 × 3.5 mm; fruits campanulate-obconic, 3–4.5 × 3–4 mm; pedicels 1.5–10 mm; peduncles 8–15 mm; venation fairly close to open; valves 3 or 4; ironbark (as in couplet 19/2) · · · · · · · · · · ·   10. *E. crebra*

21. Peduncles flat or compressed · 22

     Peduncles terete or subangular (if not clear try both leads) · 27

22. Fruits and flowers practically sessile; fruits cylindric, 9 × 6 mm; peduncles 10–12 mm long (pedicels 0–3 mm in some forms approaching *E. robusta*) · · · · · · · · · · · · · · ·   5. *E. botryoides*

     Fruits and flowers pedicellate but pedicels often short and very thick · 23

23. Fruits ³/₄-globose with narrow opening and
thick walls · 24
Fruits various but not as above · 25
24. Bark fibrous and persistent to small branches;
fruits 11 × 10.5 mm with opening 5.5 mm
wide; pedicels short, 4–6 mm long; peduncles
flat or angular, 18–25 mm long · · · · · · · · · ·   26. *E. muelleriana*
Bark fibrous and persistent on trunk for 2–9 m
then smooth and deciduous; fruits
(6.5–)8–10.5 × 9.5–11 mm with opening
4.5–8 mm wide; pedicels 2.5–7 mm long;
peduncles 15–20 mm long · · · · · · · · · · · ·   30. *E. pilularis*
25. Fruits cylindric-urceolate, shaped like a Grecian
urn, 13–15 × 7–10.5 mm; leaves ovate-
lanceolate, acuminate with venation close
and reticulate; petioles often twisted; pedicels
3–12 cm long, peduncles 25–35 mm long · · ·   (small fruited specimens of)
18. *E. gummifera*
Fruits usually not so classically urceolate · · · · · · · · · · · · · · · · 26
26. Venation very open and oblique; fruits
pyriform, 8–9 × 8–9 mm; bark deeply
furrowed; operculum hemispherical, 1–2
mm long, 2–3 mm wide · · · · · · · · · · · · · ·   27. *E. obliqua*
Venation closer, not very oblique; leaves
cuspidate; fruits narrowly ovoid-cylindric or
cylindric-urceolate, 9–15 × 8–12 mm; bark
red-brown, soft spongy and subfibrous,
rough; operculum drawn out into a narrow
cylindric tip, 11–14 mm long · · · · · · · · · ·   35. *E. robusta*
27. Fruits characteristic, elliptic-cylindric to cylindric-
urceolate, narrowed at the rim, longitudinally
ribbed, 9–10 × 5.5–7 mm; pedicels 1–9 mm
long; peduncles 15 mm long; buds and fruit on
leafless branchlets below current leaves · · · · ·   9. *E. cladocalyx*
Fruit if cylindric then not ribbed · 28
28. Pedicels 10–20 mm long; inflorescences
pendulous · 29
Pedicels shorter, under 10 mm and usually
under 6 mm long · 31
29. Leaves long, 17–22.5(–25) × 2.3–4(–5) cm;
umbels 3-flowered; fruits campanulate, 10–13
× 10–12 mm; pedicels 13–22 mm; peduncles
15–20 mm long; bark grey, deciduous in
irregular flakes from the branches and upper
portion of the stem, persistent, thick and
subfibrous on remainder of trunk · · · · · · · ·   20. *E. longifolia*
Leaves usually shorter, 9.5–12(–17) × 1–3(–3.3)
cm · 30

30. Umbels [3]7-flowered*; bark jet-black or very
dark red, hard, deeply furrowed; fruits ovoid,
slightly urceolate, 8–11 × 7–9.5 mm; pedicels
10–20 mm long; peduncles 5–18 mm long
(stems often pruinose) · · · · · · · · · · · · · · ·   39. *E. sideroxylon*

Umbels 3-flowered; bark on upper trunk and
branches smooth, mottled white and blue,
deciduous, the fibrous rough bark persistent
at base; fruits campanulate or ovoid, 9–11 ×
9–11 mm; pedicels 10–15 mm long;
peduncles 10–14 mm long (stems often
pruinose) ·   19. *E. leucoxylon*

31. Fruits cylindric or ovoid-cylindric with narrow
opening, 7–10 × 6 mm; buds, shoots, leaves
etc. tinged or densely glaucous-white;
pedicels thick, 2–3 mm long merging with
the flowers; peduncles 13 mm long · · · · · · ·   1. *E. albens*

Fruits globose or ovoid and without other
characters combined · 32

32. Bark smooth, yellow, blotched with pale or dark
blue; fruit campanulate or ovoid, 8 × 8 mm;
pedicels thick, 5 mm long; peduncles up to
23 mm long; venation close; operculum
rounded to conical · · · · · · · · · · · · · · · ·   11. *E. diversicolor*

Bark rough, at least below; venation more open · · · · · · · · · · · · · · · · · · · 33

33. Bark rough and fibrous, persisting to the small
branches; fruit ³/₄-globose, 11 × 10.5 mm;
venation fairly open and rather obscure · · · ·   26. *E. muelleriana*

Bark rough for 1.8–9 m, then smooth and
greenish white; fruit ³/₄-globose, 6.5–10.5 ×
9.5–11.5 mm with narrow opening 4.5–8 mm
wide; venation ± open with obvious marginal
vein ·   10. *E. pilularis*

35. Fruits shallowly hemispherical, 3.5–4.5 mm
long and wide, 3–4-locular, the exserted
valves ± equalling the bowl, the fruit
resembling a lythraceous capsule; pedicels
2–5 mm long; operculum hemispherical;
bark on trunk rough, grey to grey-black, full
of wrinkles and cracks (if bark smooth see
note after species 25) · · · · · · · · · · · · · · ·   25. *E. microtheca*

Fruits turbinate, campanulate etc., not shallow
nor other characters combined · 36

36. Fruits turbinate-campanulate 5–8.5 × 4.5–8.5
mm, usually with a very oblique, raised disc;
flowers usually in umbels but rarely in
panicles ·   40. *E. tereticornis*

Fruits oblong, turbinate, or obconic-
campanulate of ± similar size but without
very oblique disc · 37

* square brackets [] throughout the *Eucalyptus* account refers to information mostly obtained
from the cultivated East African material.

37. Buds with operculum distinctly conical,
    sometimes ± rostrate; bark dark grey to grey-
    black, rough ·   38. *E. siderophloia*
    Buds with operculum hemispherical or more or
    less conical · 38
38. Leaves discolorous with margins subcrenulate
    or subundulate; venation rather open with
    obvious marginal nerve; peduncles mostly
    round; pedicels 9–10 mm long; fruits oblong
    or turbinate, 6–9 × 4.5–5.5 mm; valves usually
    3, sometimes 4, thin, flush or just exsert (can
    be included); bark rough, fibrous, brown to
    red-brown ·   24. *E. microcorys*
    Leaves concolorous with other characters less
    obvious; peduncles angled or flattened;
    pedicels 3–6 mm long; fruits obconic-
    campanulate, 6 × 5.5 mm; valves 4–5, flush to
    exserted (can be included); ironbark, grey to
    black ·   12. *E. drepanophylla*
                                                                   (A few specimens of
                                                                   E. paniculata with flush
                                                                   valves could key here)
39. Fruits exceeding 10 mm in at least one direction · · · · · · · · · · · · · · · · 40
    Fruits not exceeding 10 mm in any direction · · · · · · · · · · · · · · · · · · 43
40. Fruits characteristic, campanulate, bicostate,
    8–18 × 10–19 mm with oblique rim and a
    flange 2.5–4 mm tall and valves strongly
    exsert; pedicels thick, 2–12 mm long;
    peduncles flat, ± 12 mm long; bark rough,
    fibrous and persistent · · · · · · · · · · · · · · · ·   29. *E. pellita*
    Fruits not so characteristically flanged and
    without other characters combined · 41
41. Fruits 11 × 10.5 mm with opening 5.5 mm wide;
    pedicels stout, 4–6 mm long; peduncles flat
    or angular 18–25 mm long; bark fibrous and
    persistent even on small branches · · · · · · · ·   26. *E. muelleriana*
    Without these characters combined · 42
42. Leaves concolorous, lanceolate-falcate, 19–27 ×
    1.4–2 cm (sometimes much longer), rugulose;
    venation ± open; stems rugulose; fruits
    turbinate, 10–11 × 9.5–11 mm, rim oblique;
    peduncles flat, 8–30 × 5.5 mm · · · · · · · · · · ·   22. *E. maidenii*
    Leaves discolorous, lanceolate but broader,
    13–14 × 2.3–3.5 cm; fruits cylindric-turbinate,
    6–10 × 7.5–10 mm; peduncles 12–18 mm
    long, angular or rounded · · · · · · · · · · · · ·   32. *E. punctata*
43. Venation very oblique, the secondary nerves
    almost parallel to the midrib · 44
    Venation not so markedly oblique · 45

44. Fibrous bark persistent over whole trunk; fruits turbinate, 6–7 × 5.5 mm, valves exsert; pedicels (1–)3–5 mm long; peduncles 4–14 mm long · 13. *E. fastigata*

Fibrous bark sometimes persistent on trunk for 2–8 m only and then smooth, white; fruits ± turbinate, 7 × 6–7 mm, valves just exsert (but can be included); pedicels 3–5 mm long; peduncles 10 mm long · · · · · · · · · · · · · · · 33. *E. regnans*

45. Fruits campanulate–shallow-bowl-shaped with laterally projecting rim and very exsert valves, 4.5 × 8.5 mm; pedicels 3–7 mm long; peduncles round, 8–14 mm long; operculum shortly conic · *E. camaldulensis* hybrids and some *camaldulensis* itself

Fruits without such a markedly projecting rim (or if fairly marked and operculum long-conic see *tereticornis*) · 46

46. Fruits hemispherical to turbinate with disc very obviously raised and obliquely convex, often dome-like with valves stout and strongly exsert · 47

Fruits with disc not or not very raised, truncate; valves strongly exsert to flush, or if rim raised and dome-like then valves not strongly exsert · 55

47. Bark smooth, often deciduous or sometimes a little rough persistent bark for first metre or so · 48
Bark rough and persistent, fibrous or stringy or soft box-like and persistent · 53

48. Operculum shortly conic, rounded, scarcely longer than, if as long as calyx-tube · 49
Operculum mostly longer or much longer than calyx-tube · 50

49. Fruits campanulate or subglobose, 5 × 6.5 mm; pedicels 1–4 mm long; peduncles angular, 7–12 mm long; 3-flowered · · · · · · · · · · · · 41. *E. viminalis*
Fruits half-globular, 4–7 × 4–7 mm; pedicels 5–14 mm long; peduncles round, 5–25 mm long; more than 3-flowered · · · · · · · · · · · · try 7. *E. camaldulensis*

50. Fruits turbinate-conic, 5–8 × 7.5–8 mm, sometimes with oblique rim; bark smooth or with fibrous stocking; pedicels 2–8 mm long, thick; peduncles round or flat, 7–8 mm long; fruits often greenish purple · · · · · · · · · · · · try 7. *E. camaldulensis* × *grandis/saligna* hybrids

Fruits $\frac{1}{2}$-globular to turbinate-globular or campanulate · 51

51. Operculum typically rostrate or cuspidate from
    a broad base (but not always); young leaves
    lanceolate; fruits half-globular, 4–7 × 4–7
    mm; disc oblique or truncate; rim sometimes
    flanged; pedicels 5–14 mm long; seed yellow;
    bark smooth, dull white or grey, deciduous
    throughout (umbels sometimes abnormally
    massed in spherical clusters) (by far the most
    commonly cultivated species) · · · · · · · · · ·   7. *E. camaldulensis*
    Operculum elongate conical or, if somewhat
    rostrate, then bark black decorticating in
    broad ribbons and young leaves ovate or
    round; seed black; bark smooth, blotched or
    mottled, deciduous · 52
52. Young leaves broadly lanceolate or elliptic;
    fruits 5–8.5 × 4.5–8.5 mm; disc domed with
    valves fused to it; operculum usually 2–3
    times the length of the calyx-tube; pedicels
    2–10 mm long; seed black; peduncles round
    or angular or rarely flattened, 8–15 mm long   40. *E. tereticornis*
    Young leaves ovate or orbicular; fruits 5–6 × 7
    mm; disc slightly domed; operculum 1.5–2.5
    times the length of the calyx-tube; pedicels
    angular, 5–6 mm long; peduncles ± round, 10
    mm long ·   2. *E. blakelyi*
53. Peduncles   round;   fruits   globose   or
    hemispherical, often with flanged rim, 6 × 7
    mm; leaves concolorous · · · · · · · · · · · · · ·   36. *E. rudis*
    Peduncles mostly compressed, flat or only
    angular (rarely rounded in some hybrids);
    leaves ± discolorous · 54
54. Fruits ³/₄-globose-turbinate, 6–7 × 6–7 mm;
    peduncles flat, angled or rarely rounded;
    operculum conic or long-conic · · · · · · · · · ·   34. *E. resinifera*
    Fruits turbinate, 5–8 × 7–9 mm, conic or
    campanulate, often flanged; peduncles flat or
    rarely rounded; operculum usually short but
    rostrate ·   try hybrids of *E. camaldulensis*
                                                              × *grandis/saligna*
55. Fruits truly obconic with straight sides, 7 × 7
    mm, ± sessile; valves strongly exsert; gum
    bark ·   try *E. saligna* × *grandis* and
                                                              possibly hybrids with
                                                              camaldulensis

    Fruits turbinate to hemispherical · 56
56. Peduncles round or very slender · 57
    Peduncles angled or flat · 58

57. Fruits hemispherical, 4–7 × 4–7 mm, disc
    usually domed, rarely truncate, with very
    exserted valves; pedicels 5–14 mm long;
    operculum pointed, rostrate; venation ±
    open; bark smooth and deciduous · · · · · · ·   7. *E. camaldulensis*

    Fruits turbinate, turbinate-conic, turbinate-
    hemispherical or campanulate, 5–8 × 7–9 mm,
    greenish or purplish; disc truncate or
    oblique; valves exsert; pedicels thick, 2–8 mm
    long; peduncles 7–18 mm long, round or flat;
    operculum shortly rostrate or cuspidate or
    blunt; venation open or close; bark smooth
    or with some fissures or with fibrous stocking   try *E. camaldulensis* ×
                                                            grandis/saligna and other
                                                            hybrids of *grandis/saligna*;
                                                            11. *E. diversicolor* might also
                                                            key here

58. Ironbark; bark hard, fibrous, flaky, furrowed
    and usually persistent (at least on the trunk)
    [fruits turbinate, (4–)7.5–8 × (4.5–)6.5–7.5
    mm, ± 4-ribbed, rugose or smooth; valves
    exsert; pedicels 0–4 mm long, thick;
    peduncles angular, 14–23 mm long; young
    leaves round] (flowers are usually in
    panicles) ·   38. *E. siderophloia*

    Not an ironbark, bark smooth or with some
    fissures and fibrous stocking in some hybrids,
    or part at base may be rough [note if bark is
    all flaky furrowed and persistent but other
    characters do not agree with the first lead
    above try hybrids of *E. grandis* and *E.
    tereticornis* also some forms of *E. resinifera, E.
    rudis* and *E. botryoides* and its hybrids] · 59

59. Leaves concolorous; venation fine, the veins
    narrow and delicate; inflorescence 3-
    flowered; fruits 7.5 × 8.5 mm, sessile or
    pedicels very short, 1–2 mm long · · · · · · · ·   41. *E. viminalis* forms

    Leaves discolorous; inflorescence more than 3-
    flowered · 60

60. Bark rough at base, even-coloured above; fruits
    mostly dark, usually purplish, often with a
    distinct bloom · 61

    Bark smooth mottled, finally granular; fruits
    mostly more greenish or brownish · 62

61 Buds glaucous, distinctly pedicellate, slightly
    contracted in the middle; operculum usually
    shorter than the calyx-tube; fruits slightly
    glaucous, cylindric-conic, 7–8 × 5–5.5 mm,
    rim thin; pedicels 0.5 mm long; peduncles 12
    mm long; valves 4–5, incurved; bark smooth
    and deciduous, white or subglaucous (E
    African material) · · · · · · · · · · · · · · · · · ·   17. *E. grandis*
    Buds not glaucous, shortly pedicellate;
    operculum acute, shorter than calyx-tube;
    fruits campanulate, 6–8 × 5–8 mm, the valves
    3–4, erect; pedicels 3–8 mm long; peduncles
    11–22 mm long; bark smooth and blue,
    rough and flaky at base only (E African
    material) ·   37. *E. saligna*
62. Fruits turbinate-cylindric 6–9.5 × 7.5–10 mm
    with broad exserted valves; pedicels 3–10 mm
    long; peduncles 12–18 mm long; bark
    eventually greyish, ± thick, partly deciduous,
    reddish above · · · · · · · · · · · · · · · · · ·   32. *E. punctata*
    Fruits campanulate or campanulate-urceolate,
    8 × 8 mm, with truncate rim and thin not
    broad included, flush or exserted valves;
    pedicels thick, 5 mm long; peduncles 20–25
    mm long; bark usually patchy orange-yellow
    bronze or white and smooth · · · · · · · · · · ·   11. *E. diversicolor*

KEY TO *EUCALYPTUS* SPECIES USED MOSTLY FOR ORNAMENT

Species marked with © are also placed in *Corymbia* K.D. Hill & L.A.S. Johnson

1. Flowers solitary in leaf-axils; fruits shallowly bowl-shaped or
    hemispherical, 3–5 × 5–9 cm · · · · · · · · · · · · · · · · · ·   *E. macrocarpa*
    Flowers not solitary · 2
2. Flowers in panicles · 3
    Flowers in umbels · 5
3. Stems and leaves dark brown hairy or scabrid · · · · · · · · ·   © *E. torelliana*
    Stems and leaves not hairy · 4
4. Leaves strongly lemon-scented · · · · · · · · · · · · · · · · · ·   © 7. *E. citriodora*
    Leaves not lemon-scented ·   © 21. *E. maculata*
5. Flowers large, in 3-flowered umbels; calyx-tube 1.5–2 cm
    long, strongly ribbed; peduncle broadly flattened, 7 mm
    wide; operculum scarlet with cross-like ridges · · · · · · · ·   *E. erythrocorys*
    Calyx-tube usually smaller; peduncle much narrower · · · · · · · · · · · · · · · 6
6. Operculum and calyx-tube plicate-ribbed at least in part;
    operculum strongly acuminate · 7
    Operculum and calyx-tube not plicate-ribbed · · · · · · · · · · · · · · · · · · 8
7. Buds and/or fruits glaucous; calyx-tube obconic, plicate-
    ribbed throughout ·   *E. lesouefii*
    Buds and/or fruits not glaucous; calyx-tube cylindrical,
    plicate-ribbed at base, smooth or striate above · · · · · · ·   *E. torquata*
8. Fruits 2–3.5 × 1.5–3 cm, ± contracted at orifice; flowers red · · · · · · · · · · · · 9
    Fruits much smaller · 10

9. Operculum dome-shaped, 3 mm long with short apiculus;
fruit 2–3.5 × 2–3 cm (commonly planted); flowers in 7s,
erect ·   14. *E. ficifolia*
Operculum conical or rostrate, ± 8 mm long; flowers in 3s,
pendulous ·   *E. caesia*
10. Bark smooth and deciduous; operculum usually rostrate but
sometimes only conical; fruit hemispherical, 4–7 mm long
and wide with broad ascending disc and 3–5 very exserted
valves; leaves concolorous (the commonest eucalypt in
cultivation) ·   7. *E. camaldulensis*
Bark smooth and bluish save for base of trunk where it is
rough and fibrous; operculum conical; fruit cylindrical or
campanulate, 5–8 × 4–7 mm with narrow descending disc
and 3–4 exserted valves; leaves discolorous · · · · · · · · ·   37. *E. saligna*

1. **E. albens** *Benth.*, Fl. Austral. 3: 219 (1867); Fl. Austral. 19: 394, fig. 102/i–j (1988). Types: Australia, New South Wales, Macquarie R., *Cunningham* 198 (BM!, K!, syn.) & New South Wales, New England, *Stuart* s.n. (K!, MEL, syn.) & New South Wales, between Alfords and the Range, *Leichhardt* s.n. (MEL, syn.) & Victoria, between Ten Mile Creek and Broken R., *F. Mueller* s.n. (K!, MEL, syn.)

Tree to 25 m with rough fibrous bark on trunk but white and smooth above. Juvenile leaves ovate or ± round; adult lanceolate, 10–16 cm long, 1.7–3 cm wide; petiole not or slightly flattened, 1.5–2.2 cm long. Umbels 7-flowered; peduncle quadrangular, 1–1.5 cm long, pedicels 0–5 mm long. Buds cylindrical, glaucous; operculum conical, 4–8 mm long, 4–7 mm wide. Calyx-tube 4–8 mm long, 4–7 mm wide. Fruits glaucous, cylindrical to suburceolate, 6–15 mm long, 5–10 mm wide, often ribbed. Fig. 5/1, p. 30.

UGANDA. Bunyoro District: Masindi Township, Mar. 1950, *Dale* 801!
KENYA. Nairobi Arboretum, 1961, *Brown* 3; Kisumu District: Lumbwa, *Nottidge* 2 (fide EA)
DISTR. **U** 2; **K** 3–5

2. **E. blakelyi** *Maiden*, Crit. Rev. Gen. *Eucalyptus* 4: 43 (1917); Fl. Austral. 19: 325, fig. 91/i–j (1988). Type: Australia, New South Wales, 16–24 km from Coonabarabran towards Rocky Glen, *Jensen* 129 (NSW, holo.)

Tree to 25 m with smooth patchy bark. Juvenile leaves ovate or ± rounded; adult lanceolate, 9–16 cm long, 1–2 cm wide; petiole ± terete, 1.5–2.2 cm long. Umbels 7–11-flowered; peduncle terete or angular, [0.5]* 0.7–1.9 cm long; pedicels [angular] 1–10 mm long. Operculum conical, 5–8 mm long, 3–5 mm wide. Calyx-tube hemispherical, 3–4 mm long, 3–5 mm wide. Fruits ovoid, ± globose or hemispherical, 4–7 mm long, 4–8 mm wide with broad ascending disc and 3–4 exserted valves; seed black. Fig. 5/2, p. 30

KENYA. Kisumu–Londiani District: Londiani, compt. Mt Blackett 3b, 26 Nov. 1957, *Forest Training School* 9!
DISTR. **K** 5; **T** 3

3. **E. bicostata** *Maiden, Blakely & J.H. Simmonds*, Trees Shelter & Timber New Zealand, Eucalypts: 133, t. 48/a, b, c, f, g (1929); Maiden, Crit. Rev. Gen. *Eucalyptus* 8(1): 24 (1929). Type: Australia, New South Wales, Mundaroo State Forest, Tumbarumba, *de Beuzeville* s.n. (NSW, holo.)

* square brackets [] throughout the *Eucalyptus* account refers to information mostly obtained from the cultivated East African material.

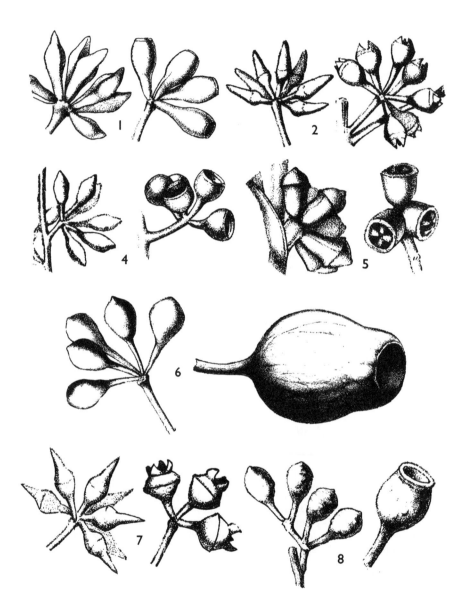

FIG. 5. Buds and fruits of *EUCALYPTUS* species, all natural size; species number as in main text. **1**, *E. albens*; **2**, *E. blakelyi*; **4**, *E. bosistoana*; **5**, *E. botryoides*; **6**, *E. calophylla*; **7**, *E. camaldulensis*; **8**, *E. citriodora*. Illustrations reproduced courtesy of the former Australian Forestry & Timber Bureau, now CSIRO Forestry & Forest Products.

Tree 12–45 m with smooth bluish bark peeling in strips or flakes. Juvenile leaves opposite, sessile, ovate-cordate to broadly lanceolate, large and glaucous; adult leaves falcate-lanceolate, 14–25 cm long, 2–3 cm wide; petiole terete or channelled, 3–5 cm long. Umbels 3-flowered; peduncle 1–3 mm long; pedicel absent or nearly so. Buds turbinate to obconical, warty, glaucous; operculum 6–8 mm long, 12–14 mm wide. Calyx-tube obconical, 7–9 mm long, 12–14 mm wide. Fruits obconic, turbinate or subglobose, glaucous, 8–17 mm long, 10–20 mm wide, 2-ribbed with 3–4 flush valves.

KENYA. Machakos District: Kalama Location, Kiitini Village, Kamutonga, 6 Dec. 1970, *Mwanthi* 1!
TANZANIA. Lushoto District: Amani (recorded by Verdcourt 1963)
DISTR. **K** 4; **T** 3

SYN. *E. globulus* Labill. var. *bicostata* (Maiden, Blakely & J.H. Simmonds) Ewart, Fl. Victoria: 804 (1931)
      E. globulus Labill. subsp. *bicostata* (Maiden, Blakely & J.H. Simmonds) J.B. Kirkp. in J.L.S. 69: 101 (1975); Fl. Austral. 19: 353, fig. 95/i–j (1988)

4. **E. bosistoana** *F. Muell.* in Austral. J. Pharm. 10: 293 (1895); Dale, Introd. Trees Uganda: 35 (1953); Fl. Austral. 19: 398, fig. 103/g–h (1988); Friis in Fl. Eth. 2 (2): 104, fig. 72.49 (1996). Type: Australia, Victoria, between Nicholson R. and Tambo R., *Schlipalius* s.n. (MEL, lecto. fide Willis in Muelleria 1: 165–166 (1967))

Tree to 60 m with rough finely fibrous bark on lower trunk or ± smooth throughout. Juvenile leaves ovate or round; adult lanceolate, 10–20 cm long, 0.7–2 cm wide, sometimes falcate; petiole terete, 1–1.7 cm long. Umbels 7-flowered; peduncle terete or quadrangular, [6]7–10 mm long; pedicels 3–10 mm long. Buds ovate to clavate; operculum conical to hemispherical, 3–5 mm long and wide. Calyx-tube hemispherical, ± 4 mm long, 3–5 mm wide. Fruits hemispherical to ovoid, 4–7 mm long and wide with moderately broad descending disc and 5–7 flush or included valves. Fig. 5/4.

UGANDA. Introduced into Ankole in 1930
KENYA. Uasin Gishu District: Kaptagat, 13 Nov. 1953, *Pudden* 14! (in part)
DISTR. **U** 2; **K** 3, 5

NOTE. A good timber used for heavy construction.

5. **E. botryoides** *Sm.* in Trans. Linn. Soc. London 3: 286 (1797); T.T.C.L.: 372 (1949); Dale, Introd. Trees Uganda: 35 (1953); F. White in F.Z. 4: 209, t. 46/e (1978); Fl. Austral. 19: 200, fig. 64/i–j (1988); Friis in Fl. Eth. 2 (2): 92, 72.6/9–10 (1996). Type: Australia, New South Wales, Port Jackson, *J. White* s.n. (LINN, holo., BM!, G, iso.)

Tree to 40 m with fibrous or flaky-fibrous bark on trunk and main branches but smooth on smaller branches. Juvenile leaves ovate; adult discolorous, broadly lanceolate, 10–16 cm long, 2.5–4 cm wide; petiole 2–3 cm long. Umbels 7–11-flowered; [very young inflorescences enclosed in bracts looking like one flower]; peduncle broadly flattened, 0.7–1.5 cm long; pedicels usually absent, less often up to 3 mm long. Buds subcylindrical, clavate or ovoid; operculum conical or hemispherical, 3–5 mm long, 4–5 mm wide. Calyx-tube cylindrical or obconical, 4–6 mm long, 4–5 mm wide, often ribbed. Fruits cylindrical, 7–12 mm long, 5–9 mm wide with moderately broad descending disc and 3–4 flush or included valves. Fig. 5/5.

UGANDA. Mubende District: Mubende Hill, Oct. 1952, *Dale* U808!, U809!, U811!
TANZANIA. Lushoto District: Amani, Drackenberg Plantation 1, 27 July 1930, *Greenway* 2334!; Iringa, Iringa College, 6 May 1972, *Pedersen* 1085 (fide EA)
DISTR. **U** 4; **K** 4, 5; **T** 1, 3, 5

NOTE. *Dale* U809 showing the bracteate inflorescences compares exactly with a Mueller sheet from Victoria named by Bentham. A good many specimens have been determined as *E. botryoides* × *E. grandis* and can have slightly exserted valves. These species are not sympatric in Australia but are often grown together in Africa.

6. **E. calophylla** *Lindl.* in Bot. Reg. 27, Misc.: 72 (1841); Fl. Austral. 19: 97, fig. 46/e–f (1988). Type: Western Australia, Albany, near Princess Royal Harbour, *R. Brown* s.n. (BM!, holo.)

Tree mostly to 40(–60) m but sometimes shrubby, with tessellated bark throughout. Juvenile leaves alternate, peltate, ovate; adult strongly discolorous, ovate to broadly lanceolate, 9–14 cm long, 2.5–4 cm wide, with conspicuous oil glands; petiole flattened, 1.5–2 cm long. Umbels usually in terminal panicles, 3–7-flowered; cream (rarely pink) flowers; peduncle terete or angular, 1.5–3.5[3.7] cm; pedicels 1–3 cm long. Buds clavate; operculum flattened-hemispherical, 2–4 mm long, 7–8 mm wide, apiculate. Calyx-tube urceolate, 5–10 mm long, 7–10 mm wide. Fruits urceolate, usually contracted at mouth, [2.7]3–5 cm long, [2.7]2.8–4 cm wide. Seeds black, not winged. Fig. 5/6, p. 30.

KENYA. Nairobi Arboretum, *F.D.* 16068; Kiambu District: Kikuyu Forest Station, *F.D.* 16067 (fide EA)
TANZANIA. Lushoto, 10 Feb. 1964, *Semsei* 3644!
DISTR. **K** 3, 4; **T** 2, 3

NOTE. Alternative name: *Corymbia calophylla* (Lindl.) K.D. Hill & L.A.S. Johnson.

7. **E. camaldulensis** *Dehnh.*, Cat. Pl. Hort. Camaldulensis ed. 2, 6: 20 (1832); T.T.C.L.: 371 (1949); Dale, Introd. Trees Uganda: 35 (1953); F. White in F.Z. 4: 210, t. 47/b (1978); Fl. Austral. 19: 327, fig. 91/s–t (1988); Thulin & G. Moggi in Fl. Somalia 1: 245 (1993); Friis in Fl. Eth. 2 (2): 98, fig. 72.7/3–4 (1996). Type: From tree cultivated in Italy, Naples, Camalduli, *Dehnhardt* s.n. (W, holo.; photo.! in Proc. Linn. Soc. New S. Wales, t.4/2 (1937))

Tree 20(–45) m tall with smooth bark. Juvenile leaves ovate to broadly lanceolate; adult lanceolate, 8–30 cm long, 0.7–2 cm wide; petiole terete or channelled, 1.2–1.5 cm long. Umbels 7–11-flowered; peduncle slender, terete or quadrangular, 0.6–1.5[2.5] cm long; pedicels slender, 0.5–1.2[1.4] cm long. Buds globular-rostrate or ovoid-conical; operculum hemispherical and rostrate to conical and obtuse, 4–6 mm long, 3–6 mm wide. Calyx-tube hemispherical, 2–3 mm long, 3–6 mm wide. Fruits hemispherical or ovoid, [4]5–8 mm long and wide with broad ascending disc and 3–5 exserted valves; seed yellow, smooth. Fig. 5/7, p. 30 & 11/1–2, p. 51.

UGANDA. West Nile District: Arua, June 1950, *Dale* U797!
KENYA. Nairobi, west of Ngara Estate, 16 Dec. 1971, *Mwangangi* 1906! & Lavington, 31 Oct. 1976, *Timberlake* 918!
TANZANIA. Lushoto District: Lushoto, near Silviculture Office, 3 June 1970, *Lysholm* 5!; R.O. Williams (U.O.P.Z: 248 (1949)) mentions this species (as *E. rostrata*) from Selem and Ziwani Plantations, Zanzibar.
DISTR. **U** 1, 3; **K** 3–5; **T** 1–7; **Z**

SYN. *E. rostrata* Schltdl. in Linnaea 20: 655 (1847), *nom. illegit., non* Cav. (1797); U.O.P.Z: 248 (1949). Type: Australia, locality and collector unknown (?HAL, type)

NOTE. Very widely grown in East Africa, this tree produces a hard durable timber used for a wide variety of purposes. One of the most widespread species in Australia and extremely variable. It readily hybridises with other species such as *E. grandis, E. saligna, E. rudis* etc. but many of the specimens determined as hybrids would seem easily accommodated within the natural variability of the species itself. Occasionally the inflorescences are abnormally crowded into heads 3 cm in diameter or more.
   E. camaldulensis is very variable in Australia and it is sometimes possible to indicate the origin of cultivated material; *Timberlake* 918 (cited above) is for example the South Australian form.

8. **E. citriodora** *Hook.* in T.L. Mitch., J. Exped. Trop. Austral.: 235 (1848); U.O.P.Z.: 248 (1949); Dale, Introd. Trees Uganda: 35 (1953); Jex-Blake, Gard. E. Afr. ed. 4: 243 (1957); F. White in F.Z. 4: 207, t. 46/b (1978); Fl. Austral. 19: 106, fig. 48/e–f (1988); Friis in Fl. Eth. 2 (2): 86, fig. 72.6/3–4 (1996). Type: Australia, Queensland, Balmy Creek, *Mitchell* 153 (K!, holo., CGE, MEL, iso.)

Tree 25–40(–50) m tall with [orange-brown] powdery [smooth] bark. Juvenile leaves ovate to broadly lanceolate, sometimes peltate, occasionally setose with long oil glands; adult lanceolate, 8–16 cm long, 0.5–2 cm wide, very strongly lemon-scented when crushed; petiole flattened, 1.3–2 cm long. Umbels 3-flowered in axillary panicles; peduncle terete, 3–7 mm long; pedicels 1–6 mm long. Buds clavate; operculum hemispherical, 3–4 mm long, 4–5 mm wide, apiculate. Calyx-tube hemispherical, 5–6 mm long, 4–5 mm wide. Fruits ovoid or urceolate, 0.7–1.5 cm long, 0.7–1.1 cm wide, often warty. Fig. 5/8, p. 30.

UGANDA. Bunyoro District: Near Budongo Forest Reserve, Nyabeya [Nyabyeya] Arboretum, 4 Oct. 1962, *Styles* 126!
KENYA. Nairobi District: Karura Forest Reserve, compt 15A, 16 Dec. 1952, *Greenway* 8763!
TANZANIA. Lushoto District: Lushoto Arboretum, 5 Apr. 1967, *Semsei* 4214!; U.O.P.Z.: 248 (1949) mentions this species as growing in the Maziaina Plantations, Zanzibar.
DISTR. **U** 2; **K** 3–5; **T** 1–4; **Z**

NOTE. Alternative name: *Corymbia citriodora* (Hook.) K.D. Hill & L.A.S. Johnson.

9. **E. cladocalyx** *F. Muell.* in Linnaea 25: 388 (1853); Fl. Austral. 19: 252, fig. 76/u–v (1988); Friis in Fl. Eth. 2 (2): 96, fig. 72.29 (1996). Type: S Australia, Marble Range, *Wilhelmi* s.n. (MEL, holo., K!, iso.)

Tree to 35 m with smooth pale bark. Juvenile leaves alternate, round; adult strongly discolorous, lanceolate, 11–15 cm long, 2–2.5 cm wide; petiole quadrangular, 1.2–2.1 cm long. Umbels 7–11-flowered, on leafless branchlets below current leaves; peduncle terete, 1–1.7 cm long; pedicels 2–7 mm long. Buds cylindrical or urceolate; operculum hemispherical, 3–4 mm long, ± 5 mm wide, apiculate. Calyx-tube cylindrical, 7–8 mm long, 4–5 mm wide, faintly ribbed. Fruits ovoid or urceolate [elliptic-cylindric to cylindric-urceolate], 9–16 mm long, 6–10 mm wide, ribbed, with broad descending disc and 3–4 included valves. Fig. 6/9, p. 35.

KENYA. Nairobi Arboretum, 24 Feb. 1953, *Darling* 20!
TANZANIA. Lushoto District: Lushoto Arboretum, 11 Nov. 1958, *Muze* 40!
DISTR. **K** 3, 4; **T** 3

10. **E. crebra** *F. Muell.* in J. Proc. Linn. Soc. Bot. 3: 87 (1859); T.T.C.L.: 371 (1949); U.O.P.Z.: 248 (1949); Dale, Introd. Trees Uganda: 36 (1953); Fl. Austral. 19: 407, figs. 105/c–d (1988); Friis in Fl. Eth. 2 (2): 104, fig. 72.50 (1996). Type: Australia, Queensland, Burdekin, *F. Mueller* s.n. (MEL, holo.)

Tree to 30 m with rough dark grey to black ironbark throughout. Juvenile leaves linear to narrowly lanceolate, adult lanceolate, 5–15 cm long, 1–1.7 cm wide; petiole 1–1.5 cm long. Umbels 7–11-flowered, arranged in panicles; peduncle terete to quadrangular, 0.4–1.2[1.5] cm long; pedicels 1–6[10] mm long with angles sometimes continuing as ribs on calyx-tube. Buds clavate or fusiform; operculum hemispherical to conical, 2–3 mm long, ± 3 mm wide. Calyx-tube hemispherical to ovoid, 2–3 mm long, ± 3 mm wide. Fruits hemispherical or ovoid, [3]4–7 mm long, [3]4–6 mm wide with narrow level or descending disc and 3–4 ± flush to included valves.

UGANDA. Ankole fide Dale

KENYA. Nakuru District: Menengai, 20 Aug. 1953, *Dillon* 3 (fide EA); Kiambu District: Muguga
  Arboretum, 28 June 1963, *Verdcourt* 3669E!
TANZANIA. Lushoto District: Lushoto Arboretum plot no. 126, 3 Mar. 1976, *Shabani* 1117!;
  (U.O.P.Z.: 248 (1949)) mentions a tree growing in the Migombani Gardens, Zanzibar
DISTR. **U** 2; **K** 3, 4; **T** 3, 6; **Z**

SYN. [*E. racemosa* sensu Blakely, Key to the Eucalypts ed. 2: 260 (1955), *non* Cav.]

NOTE. Produces an extremely durable hard strong timber much used for heavy construction,
  railway sleepers etc.

11. **E. diversicolor** *F. Muell.*, Fragm. 3: 131 (1863); T.T.C.L.: 372 (1949); Fl. Austral.
19: 196, fig. 64/a–b, photo. 18 (1988); Friis in Fl. Eth. 2 (2): 90, fig. 72.18 (1996).
Type: W Australia, ?Wilson Inlet, *Oldfield* 788 (MEL, holo., CGE, E, K!, iso.)

Tree to about 90 m with usually patchy orange-yellow, bronze or white smooth bark
throughout. Juvenile leaves broadly ovate to round; adult discolorous, broadly
lanceolate, 9–12 cm long, 2–3.2 cm wide; petiole channelled, 1–2 cm long. Umbels
7-flowered; peduncle flattened or angular (± terete), 1.8–2.8 cm long; pedicels 3–6
mm long. Buds clavate; operculum rounded, conical, 5–7 mm long and wide. Calyx-
tube cylindrical to obconical, 7–8 mm long, 5–7 mm wide. Fruits ovoid or
subglobular, 8–12 mm long, 7–10 mm wide with broad descending disc and 3
prominent flush [slightly exserted] or included valves. Fig. 6/11.

KENYA. Kisumu–Londiani District: Londiani, Mt Blackett, 26 Nov. 1957, *Forestry Training School* 8!;
  Masai District: Ngong, 23 Feb. 1953, *Darling* 10 (fide EA)
TANZANIA. Lushoto District: Amani, Drackenberg Plantation 1, 29 July 1930, *Greenway* 2321!
DISTR. **K** 3–6; **T** 3

NOTE. Karri is one of the tallest trees and yields a heavy strong durable timber obtainable in
  longer lengths than is possible from other hardwood species. Some material from Lushoto,
  Hazina, near Maguzoni road, 15 Nov. 1971, *Magogo* 181 is tentatively suggested to be this
  species, but is not. Ken Hill has suggested it might not even be a *Eucalyptus*. It is unfortunately
  only in bud.

12. **E. drepanophylla** *Benth.*, Fl. Austral. 3: 221 (1867); T.T.C.L.: 371 (1949); Fl.
Austral. 19: 407, fig. 105/a–b (1988). Type: Australia, Queensland, Port Denison,
Edgecombe Bay, *Dallachy* s.n. (NSW, lecto., BRI, K!, MEL, isolecto.)

Tree to 30 m with ironbark on trunk and larger branches or throughout, dark grey
to black and rough. Juvenile leaves elliptic to ovate, adult lanceolate, 10–15 cm long,
1.4–2.5 cm wide; petiole 1–2 cm long. Umbels 7-flowered, arranged in panicles;
peduncle terete or rarely quadrangular, 4–7 mm long; pedicels 3–6 mm long. Buds
obovoid or obconical; operculum hemispherical-conic, ± 3 mm long, ± 4 mm wide,
rounded at apex. Calyx-tube obconical, 3–4 mm long, ± 4 mm wide. Fruits
hemispherical to subcylindrical [obconic-campanulate], 4–6 mm long, 5–6 mm wide
with narrow descending disc and usually 4 flush or exserted valves [sometimes
included]. Fig. 6/12.

KENYA. Nairobi District: Karura Forest Reserve, compt 15A, 420 (1909), 16 Dec. 1952, *Greenway*
  8759!
TANZANIA. Lushoto District: Amani, Kiumba Plot 10, 27 Jan. 1931, *Greenway* 2868!
DISTR. **K** 3, 4; **T** 2, 3

13. **E. fastigata** *H. Deane & Maiden* in Proc. Linn. Soc. New S. Wales 21: 809
(1897); Fl. Austral. 19: 159, fig. 58/e–f (1988); Friis in Fl. Eth. 2 (2): 88, fig. 72.13
(1996). Type: Australia, New South Wales, Tantawanglo Mt, *Deane & Maiden* s.n.
(NSW, holo., K!, iso).

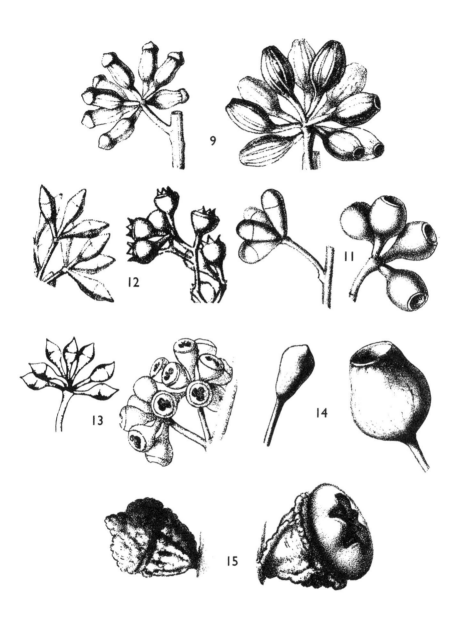

Fig. 6. Buds and fruits of *EUCALYPTUS* species, all natural size; species number as in main text.
9, *E. cladocalyx*; **11**, *E. diversicolor*; **12**, *E. drepanophylla*; **13**, *E. fastigata*; **14**, *E. ficifolia*; **15**, *E. globulus*. Illustrations reproduced courtesy of the former Australian Forestry & Timber Bureau, now CSIRO Forestry & Forest Products.

Tree to 45 m with rough furrowed bark on trunk and main branches, shedding in long strips above so that upper branches are smooth and white. Juvenile leaves broadly lanceolate to ovate, oblique; adult lanceolate, 8–15 cm long, 1.5–2.7 cm wide, [with very open oblique venation]; petiole flattened or channelled, 1–1.5 cm long. Umbels 11–15-flowered, often paired in leaf axils; peduncle terete or angular, 0.4–1.4 cm long; pedicels 1–2[3–5] mm long. Buds clavate; operculum conical or rostrate, ± 2 mm long, ± 3 mm wide. Calyx-tube obconic, ± 2 mm long, 3 mm wide. Fruits obconical to pyriform, 5–8 mm long, 4–7 mm wide, with broad, level or slightly ascending disc and 3 flush or slightly exserted valves. Fig. 6/13, p. 35.

KENYA. Uasin Gishu District: Kaptagat, *Pudden* 13!
TANZANIA. Arusha District: Olmotonyi Forestry School, Feb. 1979, *Lema* 48!
DISTR. **K** 3, 4; **T** 2, 3

14. **E. ficifolia** *F. Muell.*, Fragm. 2: 85 (1860); Dale, Introd. Trees Uganda: 36 (1953); F. White in F.Z. 4: 207, t. 46a (1978); Fl. Austral. 19: 97, fig. 46/c–d, photo. 20 (1988); Friis in Fl. Eth. 2 (2): 86, fig. 72.6/1–2 (1996). Type: Western Australia, Broken Inlet, *Maxwell* s.n. (MEL, holo., K!, iso.)

Tree to 10 m with rough fibrous bark throughout. Juvenile leaves usually alternate, sometimes peltate, ovate or round, bristly; adult strongly discolorous, alternate, rarely opposite, broadly lanceolate to ovate, 7.5–15 cm long, 3–5 cm wide, without conspicuous oil glands; petiole flattened or channelled, 1–2 cm long. Umbels 3–7-flowered, usually in terminal panicles; peduncle angular, 1.5–2.5 cm long; pedicels 1.5–2.5 cm long. Flowers crimson, orange or pink, but white varieties occur in cultivation; buds clavate or obovoid; operculum depressed hemispherical-conical, 2–3 mm long, 6–7 mm wide. Calyx-tube 8–10 mm long, 6–7 mm wide. Fruits ovoid, subglobose or urceolate, usually contracted at the mouth, 2–3.5 cm long, 2–3 cm wide. Seeds brownish with terminal wing. Fig. 6/14, p. 35.

UGANDA. Dale gives Kigezi District: Kabale; Toro District: Fort Portal; Mengo District: Nakawa (Kampala)
KENYA. Nairobi Arboretum, 12 May 1952, *Williams Sangai* 399!
TANZANIA. Lushoto District: Lushoto Arboretum, 14 Dec. 1966, *Semsei* 4152!
DISTR. **U** 2, 4; **K** 3; **T** 3

NOTE. Alternative name: *Corymbia ficifolia* (F. Muell.) K.D. Hill & L.A.S. Johnson.

15. **E. globulus** *Labill.*, Voy. Rech. Pérouse 1: 153, t. 13 (1800); T.T.C.L.: 371 (1949); Dale, Introd. Trees Uganda: 36 (1953); Fl. Austral. 19: 352, fig. 95/g–h (1988); Thulin & G. Moggi in Fl. Somalia 1: 245 (1993); Friis in Fl. Eth. 2 (2): 100, fig. 72.39 (1996). Type: Australia, Tasmania, probably Recherche Bay, *Labillardière* s.n. (FI, holo., BM!, G, K!, L, W, iso.)

Tree 30–70 m tall with bark flaking off to reveal smooth white to cream, yellow or grey surface save at extreme base of trunk. Juvenile leaves opposite, sessile and amplexicaul [ovate-oblong, cordate]; adult leaves lanceolate, 12–25 cm long, 1.7–3 cm wide, thick; petiole channelled or flattened, 2–3[–3.5] cm long. Flowers solitary in leaf-axils (rarely in threes). Peduncle 0–6 mm long, compressed; pedicels absent or very short; buds turbinate to obconic, glaucous and warty; operculum flattened hemispherical, 0.7–1.5 cm long, 1.4–1.7 cm wide with prominent umbo. Calyx-tube obconical, 1–1.2 cm long, 1.4–1.7 cm wide, ribbed. Fruits obconic to hemispherical or subglobular, glaucous, 1–2.1 cm long, 1.4–2.4 cm wide with 3–5 flush or included valves; disc broad, level to ascending. Fig. 6/15, p. 35.

UGANDA. Kigezi District: Kigezi, July 1967, *Rwamushane* s.n. (fide EA).
KENYA. Kiambu District: Hoyi Forest Station, Feb. 1933, *Dale* 656 (fide EA)

TANZANIA. Lushoto District: Lushoto, 12 May 1947, *Hughes* 40!, 41!
DISTR. **U** 2; **K** 3, 4; **T** 3, 7

NOTE. The Tasmanian blue gum is one of the best known and most easily recognised eucalypts and much more widespread than the specimens suggest. The taxa often recognised as subspecies are here kept as species.

16. **E. gomphocephala** *DC.*, Prodr. 3: 220 (1828); Fl. Austral. 19: 208, fig. 66/g–h (1988); Friis in Fl. Eth. 2 (2): 93, fig. 72.22 (1996). Type: W Australia, Geographe Bay, *Leschenault* s.n. (G, holo, K!, iso.)

Tree to 40 m with grey finely fissured flaky-fibrous bark throughout. Juvenile leaves ovate, often cordate, adult lanceolate, 9–16 cm long, 1.6–2.5 cm wide; petiole flattened or channelled, 1.5–2 cm long. Umbels 7-flowered; peduncle broad and flat, 1.3–1.7[3] cm long [8 mm wide]; pedicels absent or up to 4 mm long. Buds mushroom-shaped; operculum hemispherical, distinctly wider than calyx-tube apex, 8–10 mm long, 0.9–1.3 cm wide. Calyx-tube obconical or campanulate, 7–9 mm long, 7–8 mm wide, often ribbed. Fruits campanulate or cylindrical, 1.3–2.2 cm long, 1.3–1.7 cm wide, often faintly ribbed with broad level, convex or ascending disc and 4 flush or slightly exserted valves. Fig. 7/16, p. 38.

KENYA. Central Kavirondo District: Maseno, 27 Jan. 1953, *Abraham* 6!
TANZANIA. Arusha District: Olmotonyi Forestry School, Feb. 1979, *Lema* 46!
DISTR. **K** 3–5; **T** 2

NOTE. Tuart yields a very hard strong and durable wood.

17. **E. grandis** *W. Hill**, Cat. Nat. Industr. Prod. Queensland (London Exhib.): 25 (1862) (not seen); Dale, Introd. Trees Uganda: 37 (1953); F. White in F.Z. 4: 209, t. 46/d (1978); Fl. Austral. 19: 198, fig. 64/e–f (1988); Friis in Fl. Eth. 2 (2): 91, fig. 72.6/7–8 (1996). Type: Australia, New South Wales, *Hill* 74 (K!, ? holo.)

Tree to 55 m with white, grey-white or blue-green smooth bark except for basal 4 m which has some rough flaky bark. Juvenile leaves ovate, adult lanceolate, 10–16 cm long, 2–3 cm wide; petiole channelled, 1.5–2 cm long. Umbels 7–11-flowered; peduncle flattened, 0.8–1.8 cm long; pedicels absent or angular, up to 3 mm long. Buds ovoid or broadly fusiform [glaucous]; operculum conical or slightly rostrate, 3–4 mm long, 4–5 mm wide [usually shorter than calyx-tube]. Calyx-tube obconical to campanulate, 3–4 mm long, 4–5 mm wide. Fruits subpyriform [cylindric-conic], 5–8 mm long, 4–7 mm wide with narrow level or descending disc and 4–5[6] scarcely exserted [or flush] incurved [match-like] valves [contracted at mouth]. Fig. 7/17, p. 38.

UGANDA. Mengo District: Namanve Plantations compt 13, 28 Apr. 1959, *Hughes* 352!
KENYA. Kisumu-Londiani District: Londiani Forest Nursery, 10 Feb. 1953, *Pudden* 12 (EA);
    Machakos District: 5.5 km N of Nunguni, Kithemba Village, 10 May 1968, *Mwangangi* 912!
TANZANIA. Iringa District: Lugoda tea estate, 10 May 1968, *Renvoize & Abdallah* 2091!
DISTR. **U** 1?, 3, 4; **K** 4, 5; **T** 2–4, 7

NOTE. An important timber tree with wood resistant to borers. Hybrids said to be with *E. tereticornis, E. botryoides* and *E. camaldulensis* have been reported from several places.

18. **E. gummifera** *(Gaertn.) Hochr.* in Candollea 2: 464 (1925); T.T.C.L.: 371 (1949); Fl. Austral. 19: 100, fig. 47/a–b (1988). Type: Australia, New South Wales, Botany Bay, *Banks & Solander* (BM!, holo.)

---

* See Chapman, Australian Plant Name Index (D–J): 1237 (1991).

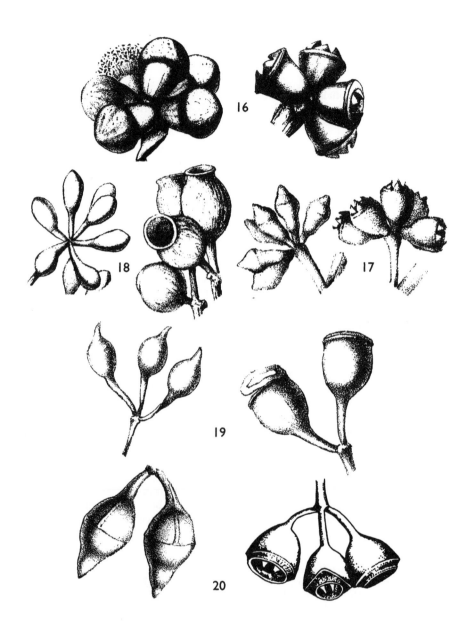

FIG. 7. Buds and fruits of *EUCALYPTUS* species, all natural size; species number as in main text. **16**, *E. gomphocephala;* **17**, *E. grandis;* **18**, *E. gummifera;* **19**, *E. leucoxylon;* **20**, *E. longifolia.* Illustrations reproduced courtesy of the former Australian Forestry & Timber Bureau, now CSIRO Forestry & Forest Products.

Tree to 35 m with tessellated rough bark throughout. Juvenile leaves alternate, ovate to broadly lanceolate, setose; adult lanceolate to broadly lanceolate, 10–14 cm long, 2–4 cm wide; petiole flattened, 1–2.3 cm long. Umbels 7-flowered, usually in terminal panicles; peduncle flattened, angular, or terete, 1.7–3[3.5] cm long; pedicels [0.3]0.9–1.4 cm long. Buds obovoid to clavate; operculum hemispherical-conical, 2–4 mm long, 6–7 mm wide, apiculate. Calyx-tube obconic, 7–9 mm long, 5–8 mm wide. Fruits exactly urceolate with distinct neck, [1.3]1.5–2 cm long, [0.7]1.1–1.5 cm wide, valves 4, very deeply included; seed unwinged. Fig. 7/18.

KENYA. Recorded from Kiambu District: Muguga and Nakuru District: Molo
TANZANIA. Lushoto District: Amani, Drackenberg Plantation 1, 29 July 1930, *Greenway* 2326!
DISTR. **K** 3, 4; **T** 3

SYN. *Metrosideros gummifera* Gaertn., Fruct. & Sem. 1: 170, t. 34, fig. 1 (1788)
    Eucalyptus corymbosa Sm., Spec. Bot. New Holl. 1: 43 (1795). Type: Australia, New South Wales, Port Jackson, *J. White* (LINN, holo.)

NOTE. Alternative name: *Corymbia gummifera* (Gaertn.) K.D. Hill & L.A.S. Johnson.

19. **E. leucoxylon** *F. Muell.* in Trans. Victorian Inst. Advancem. Sci. 1: 33–34 (1855); Dale, Introd. Trees Uganda: 37 (1953); Fl. Austral. 19: 421, fig. 108/i–j, photo. 37 (1988); Friis in Fl. Eth. 2 (2): 105, fig. 72.52 (1996). Type: S Australia, Mt Lofty range, *F. Mueller* s.n. (MEL, lecto.)

Tree to 16 m with grey to dark grey rough fibrous bark for 2 m then white, grey, yellow and/or blue smooth bark on remainder. Juvenile leaves sessile, ovate to broadly lanceolate; adult lanceolate, 9–13 cm long, 1.3–2.5 cm wide; petiole 1–2 cm long. Umbels 3-flowered; peduncle terete, 4–11[14] mm long; pedicels 0.3–2.5 cm long. Buds ovoid to fusiform, rostrate; operculum 4–5 mm long, 5–6 mm wide. Calyx-tube 5–7 mm long, 6–7 mm wide. Fruits pendulous, ovoid to subglobose, 0.7–1.4 cm long, 0.8–1.3 cm wide with dark caducous staminal ring and 4–6 included valves. Fig. 7/19.

KENYA. Recorded from Nairobi and Kisumu–Londiani District: Londiani; Kiambu, 4 Feb. 1961, *Foster* H52/61 (EA)
TANZANIA. Recorded from Lushoto District: Amani
DISTR. **K** 4, 5; **T** 3

NOTE. The yellow gum or blue gum yields very durable strong timber. The above description refers only to the nominate subspecies; three others occur in Australia. Some material said to be subsp. *megalocarpa* Boland (var. *macrocarpa* J.E. Brown) has fruits 1.5–2.2 × 1.5–1.8 cm with peduncles 1.5–3 cm long and pedicels 2–3 cm long.

20. **E. longifolia** *Link*, Enum. Hort. Berol. 2: 29 (1822); T.T.C.L.: 372 (1949); Fl. Austral. 19: 206, fig. 66/c–d (1988). Type: Australia, probably from New South Wales, Port Jackson, collector unknown (original reference merely gives Australia) (presumably B†*)

Tree to 35 m with grey ridged and cracked subfibrous bark on trunk and larger branches and bright brown or grey-green and smooth on smaller ones. Juvenile leaves ovate to broadly lanceolate; adult lanceolate, falcate, 11–24 cm long, 1.2–2.5[4] cm wide; petiole channelled, 1.5–3 cm long. Umbels 3-flowered, pendulous; peduncle terete or angular, 1.1–3.4 cm long; pedicels 0.4–2[2.2] cm long. Buds fusiform; operculum conical, 8–11 mm long, 6–12 mm wide. Calyx-tube obconical, 6–9 mm long, 6–12 mm wide. Fruits cylindrical or subcampanulate, 1–1.7 cm long, 0.9–1.6 cm wide with broad descending disc and usually 4 ± flush valves. Fig. 7/20.

---

* Fl. Austral. gives holo.: n.v.; iso.: n.v. but it is not clear what evidence there is for two specimens.

KENYA. Recorded from Nairobi and Kiambu District: Muguga
TANZANIA. Lushoto District: Amani, Drackenberg Plantation 1, 29 July 1930, *Greenway* 2320!
DISTR. **K** 4; **T** 3

21. **E. maculata** *Hook.*, Ic. Pl. 7: 619 (1844); T.T.C.L.: 371 (1949); Dale, Introd. Trees
Uganda: 37 (1953); F. White in F.Z. 4: 207, t. 46c (1978); Fl. Austral. 19: 107, fig.
48/g–h (1988); Friis in Fl. Eth. 2 (2): 87, fig. 72.6/5–6 (1996). Types: Australia, New
South Wales, Maitland, *Backhouse* 37 (K!, syn.) & New Holland, *Fraser* s.n. (K!, syn.) &
Liverpool, *collector not known* (MEL, syn.) & Newcastle, ? *Leichhardt* s.n. (MEL, syn.)

Tree to 45 m with smooth usually mottled often dimpled bark. Juvenile leaves ovate,
some peltate, at first setose; adult lanceolate, 12–21 cm long, 1.2–3 cm wide, not lemon-
scented; petiole angular, 1.5–2.5 cm long. Umbels 3-flowered in axillary panicles;
peduncles terete, [primary up to 20 mm] secondary 3–8 mm long; pedicels angular,
3–7[10] mm long. Buds ovoid; operculum hemispherical, 4–5 mm long, 5–8 mm wide.
Calyx-tube hemispherical, 5–8 mm long and wide. Fruits ovoid or suburceolate,
[0.8]1–1.4 cm long, 0.9–1.1 cm wide, valves 3–4, deeply included. Fig. 8/21.

UGANDA. Bunyoro District: Bunyoro Forest Reserve, Nyabeya [Nyabyeya] Arboretum, 3 Oct.
1962, *Styles* 113!
KENYA. Nairobi, 19 Mar. 1951, *Verdcourt* 451!
TANZANIA. Lushoto District: Amani, Kiumba Plantation 4/9, 26 Jan. 1931, *Greenway* 2858!
DISTR. **U** 1, 2, 4; **K** 1, 3, 4; **T** 1, 3–7

NOTE. A very commonly and widely cultivated species. Alternative name: *Corymbia maculata*
(Hook.) K.D. Hill & L.A.S. Johnson.

22. **E. maidenii** *F. Muell.* in Proc. Linn. Soc. New S. Wales ser. 2, 4: 1020 (1890);
Dale, Introd. Trees Uganda: 37 (1953). Type: Australia, New South Wales, Colombo,
Bäverlen s.n. (NSW, lecto., MEL, isolecto.)

Tree 15–46(–54) m tall with dark deciduous bark (save at extreme base of trunk)
leaving smooth white-yellow beneath; bluish white or blotched stems and branches.
Juvenile leaves opposite for many pairs, 4–16(–91) cm long, 4–12 cm wide; adult
lanceolate, 12–28 cm long, 1.2–2.5(–3) cm wide, thick; petiole channelled or terete,
1.5–3.5 cm long. Umbels 7-flowered; peduncle compressed, 0.8–2.5 cm long [5.5 mm
wide]; pedicels absent or up to 8 mm long. Buds glaucous; operculum hemispherical
to broadly conical, 3–4 mm long, 5–7 mm wide, smooth to distinctly verrucose. Calyx-
tube 5–7 mm long and wide. Fruits turbinate, (5–)8–11 mm long (6–)10[–12] mm
wide, 1–2-ribbed with 3–4 strongly exserted valves; disc thick, convex, partly fused to
valves. Fig. 8/22.

KENYA. Kiambu District: Muguga, 11 Mar. 1963, *Verdcourt* in EA 12800 (fide EA); Masai District:
Narok Show Ground, 14 Dec. 1963, *Verdcourt* 3842!
TANZANIA. Moshi District: Kilimanjaro, TBL estates, Osirwa Farm, 26 Jan. 1994, *Grimshaw*
94/142 & 94/140!
DISTR. **K** 4, 6; **T** 2, 3

SYN. *E. globulus* Labill. subsp. *maidenii* (F. Muell.) J.B. Kirkp. in J.L.S. 69: 101 (1974); Fl. Austral.
19: 354, fig. 95/m–n (1988); Friis in Fl. Eth. 2 (2): 101 (1996)

NOTE. Ken Hill considers that *E. bicostata* Maiden, Blakely & J.H. Simmonds and *E. maidenii*
(both considered subspecies of *E. globulus* in Fl. Austral. 19) are worthy of specific rank.
Certainly *E. globulus* and *E. maidenii* as cultivated in East Africa seem very distinct.

23. **E. melliodora** *Schauer* in Walp., Repert. Bot. Syst. 2: 924 (1843); Dale, Introd.
Trees Uganda: 38 (1953); Fl. Austral. 19: 421, fig. 108/g–h, photo. 8 (1988); Friis in
Fl. Eth. 2 (2); 105, fig. 72.51 (1996). Type: Australia, New South Wales, New Bathurst,
Cunningham 57 (ubi?, holo., E, G, K!, MEL, NSW, iso.)

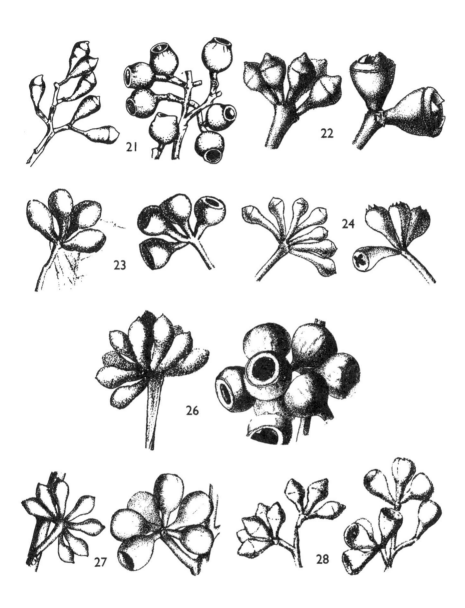

FIG. 8. Buds and fruits of *EUCALYPTUS* species, all natural size; species number as in main text. **21**, *E. maculata*; **22**, *E. maideni*; **23**, *E. melliodora*; **24**, *E. microcorys*; **26**, *E. muelleriana*; **27**, *E. obliqua*; **28**, *E. paniculata*. Illustrations reproduced courtesy of the former Australian Forestry & Timber Bureau, now CSIRO Forestry & Forest Products.

Tree to 30 m with grey, yellow or red-brown fibrous bark on lower trunk and sometimes larger branches, yellowish white and smooth above. Juvenile leaves ovate or elliptic; adult lanceolate, 6.5–14 cm long, 0.8–1.8 cm wide; petiole terete or slightly flattened, 1–1.5 cm long. Umbels 7-flowered; peduncle terete or quadrangular, 0.3–1.1[1.5] cm long; pedicels 2–9 mm long. Buds clavate to fusiform; operculum conical to rostrate, 2–3 mm long, ± 3 mm wide, usually narrower than the hemispherical calyx-tube which is 3–4 mm long and wide. Fruits hemispherical, ovoid to subglobular, 4–7 mm long and wide with narrow descending disc and obscured by a prominent dark staminal ring and usually 5 flush or included valves. Fig. 8/23, p. 41.

UGANDA. Introduced into Ankole District, Ankole, 1930
KENYA. Nakuru District: Bahati Forest Station, 20 Jan. 1953, *Pudden* 1!
DISTR. **U** 2; **K** 3–5

NOTE. Yields an extremely durable strong timber.

24. **E. microcorys** *F. Muell.*, Fragm. 2: 50 (1860); T.T.C.L.: 372 (1949); Dale, Introd. Trees Uganda: 38 (1953); Fl. Austral. 19: 426, fig. 109/g–h (1988); Friis in Fl. Eth. 2 (2): 106, fig. 72.54 (1996). Types: Australia, Queensland, Brisbane R., *F. Mueller* s.n. (K!, MEL, syn.); New South Wales, Hastings R., *Beckler* s.n. (K!, MEL, syn.) & Macleay R., *Beckler* s.n. (ubi, syn.)

Tree to 60 m with brown or red-brown rough softly fibrous bark throughout. Juvenile leaves ovate, often crenulate; adult lanceolate, 8–13 cm long, 1.5–2.5 cm wide, usually crenulate; petiole terete to channelled, 0.8–1.5 cm long. Umbels 7–9-flowered (Fl. Austral.), simple or in terminal and axillary panicles*; peduncle flattened, 0.6–1.8 cm long [round in much East African material]; pedicels 2–7[9–10] mm long. Buds clavate; operculum hemispherical, 1–2 mm long, 2–3 mm wide, often marked with crossed sutures. Calyx-tube clavate, 3–5 mm long, 2–3 mm wide. Fruits obconical, 4–10 mm long, 3–6 mm wide, usually 3-locular, with moderately broad steeply descending disc and 3–4 included to slightly exserted valves. Fig. 8/24, p. 41.

UGANDA. Busoga District: Mutai Plantations, compt 13, *Kingston* in *F.D.* 2200! & 2201!
KENYA. Kiambu District: Muguga Forest Reserve, compt 3A, 16 Dec. 1952, *Greenway* 8752!
TANZANIA. Lushoto District: Nderema [Ndarema] Road, 6 Dec. 1928, *Greenway* 1028!
DISTR. **U** 2, 3; **K** 3–5; **T** 3, 7

25. **E. microtheca** *F. Muell.* in J.L.S. 3: 87 (1859); Fl. Austral. 19: 379, fig. 100/a–b (1988); Friis in Fl. Eth. 2 (2): 103, fig. 72.47 (1996). Type: Australia, Northern Territory, Victoria R., *F. Mueller* s.n. (MEL, holo., BM!, CANB, NSW, iso.)

Tree to 20 m with bark rough throughout, grey to grey-black. Juvenile leaves lanceolate; adult lanceolate, 8–17 cm long, 0.8–2.5 cm wide; petiole terete, 0.8–1.7 cm long. Umbels 7-flowered, arranged in terminal panicles; peduncle terete, 3–9 mm long; pedicels 1–4[5] mm long. Buds usually ovoid or sometimes fusiform, often glaucous; operculum hemispherical or conical, 1–2 mm long, 2–3 mm wide. Fruits shallowly hemispherical to obconic, 1–5 mm long, 3–7 mm wide with or without a very narrow ascending disc; valves 3–4, strongly exserted.

KENYA. Nairobi Arboretum, 24 Feb. 1953, *Darling* 18!
TANZANIA. Dodoma District: Kigwe Provenance Trail, 25 June 1973, *Ruffo* 860!
DISTR. **K** 4; **T** 3, 5

---

* No mention is made that the inflorescences can be anything but simple axillary umbels in Fl. Austral. but some East African material certainly has panicles and the fig. g–h does not show simple axillary umbels.

NOTE. The coolibah or coolabah tree. I had followed Fl. Austral. in describing the bark as smooth throughout to rough throughout but Ian Brooker has pointed out that three species are involved; completely smooth-barked trees are *E. victrix* L.A.S. Johnson & K.D. Hill, and half-rough-barked trees are *E. coolabah* Blakely & Jacobs.

26. **E. muelleriana** *A.W. Howitt* in Trans. Roy. Soc. Victoria n. ser. 2: 89 (1890); Fl. Austral. 19: 142, fig. 55/k–l (1988). Type: Victoria, S Gippsland, near Hedley, Nine Mile Creek, *Howitt* 6 (MEL, lecto. fide Willis in Muellera 1: 127 (1967))

Tree to 40 m with grey-brown fibrous stringy bark throughout. Juvenile leaves ovate, often oblique; adult lanceolate, 8–13 cm long, 1.5–2.5 cm wide, distinctly unequal at the base; petiole channelled, 1.5–2 cm long. Umbels 7–11-flowered; peduncle angular or flattened, 0.8–1.8[2.5] cm long; pedicels 1–3[4–6] mm long. Buds ovoid to clavate; operculum hemispherical to conical, 2–3 mm long, 3–4 mm wide. Calyx-tube obconical, 3–5 mm long, 3–4 mm wide. Fruit subglobose to truncate-ovoid, 7–10[11] mm long, 8–12 mm wide with broad level, slightly descending or sometimes slightly convex disc and 4 slightly included or slightly exserted valves. Fig. 8/26, p. 41.

KENYA. Kiambu District: Muguga Arboretum, 28 June 1963, *Verdcourt* 3669C!
TANZANIA. Recorded from Lushoto District: Lushoto
DISTR. **K** 3, 4; **T** 3

27. **E. obliqua** *L'Hér.*, Sert. Angl.: 18 (1788), t. 20 (1792); Fl. Austral. 19: 160, fig. 58/k–l (1988); Friis in Fl. Eth. 2 (2): 89, fig. 72.15 (1996). Type: Australia, Tasmania, Adventure Bay, *Nelson* s.n. (BM!, holo.)

Tree to 90 m with grey to red-brown furrowed stringy fibrous bark throughout. Juvenile leaves ovate, oblique; adult broadly lanceolate, 10–15 cm long, 1.5–5 cm wide, shining; petiole channelled, 0.7–1.7 cm long. Umbels 11- or more-flowered; peduncles angular or flattened, 0.4–1.5 cm long; pedicels 1–6 mm long. Buds clavate; operculum hemispherical, 1–2 mm long, 2–3 mm wide, apiculate. Calyx-tube 2–4 mm long, 2–3 mm wide. Fruits ovoid, subglobose, barrel-shaped or urceolate, 6–11 mm long, 5–9 mm wide with level or steeply descending disc and 3–4 included valves. Fig. 8/27, p. 41.

KENYA. Kiambu District: Lari, *Pudden* 054! & Uplands Arboretum, 1958, *Pudden* s.n. (fide EA)
TANZANIA. Lushoto District: Amani, *A.H.* 9142 (fide T.T.C.L.)
DISTR. **K** 4; **T** 3

NOTE. This is the type species of the genus. The messmate is one of the most important Australian hardwoods.

28. **E. paniculata** *Sm.* in Trans. Linn. Soc. London 3: 287 (1797); Dale, Introd. Trees Uganda: 38 (1953); F. White in F.Z. 4: 210, t. 47/c (1978); Fl. Austral. 19: 417, fig. 107/k–l (1988); Friis in Fl. Eth. 2 (2): 105, fig. 72.7/5–6 (1996). Type: Australia, New South Wales, Port Jackson, *Burton* s.n. (LINN, holo., BM!, iso.)

Tree to 50 m with light grey rough bark throughout [deeply fissured]. Juvenile leaves ovate to broadly lanceolate; adult discolorous, lanceolate, 9.5–15 cm long, 1.2–2.4 cm wide; petiole 1.3–2.2 cm long. Umbels 7-flowered in terminal panicles and also some axillary; peduncle terete, quadrangular or flattened, 0.6–1.6[2] cm long; pedicels 4-angled, 2–10 mm long. Buds obovoid to fusiform; operculum conical, usually narrower than calyx-tube, 3–4 mm long and wide. Calyx-tube obovoid to obconical, 4–5 mm long and wide. Fruits hemispherical, obconical, obovoid or subpyriform, [5]6–8[8.5] mm long, 5–8 mm wide with obscure descending disc and 4–5 flush or included valves. Fig. 8/28, p. 41 & 11/3–4, p. 51.

UGANDA. Ankole District: between Rushasha and Rwanyampazi Kasari, 31 May 1970, *Katende* 305!
KENYA. Kiambu District: Muguga Forest Reserve, compt 3A, 16 Dec. 1952, *Greenway* 8751!
TANZANIA. Lushoto District: Korogwe District, Silviculture Nursery, 25 Jan. 1971, *Ruffo* 375!
DISTR. **U** 2; **K** 3–5, 7; **T** 1–3

29. **E. pellita** *F. Muell.*, Fragm. 4: 159 (1864); T.T.C.L.: 372 (1949); Fl. Austral. 19: 201, fig. 65/a–b (1988). Type: Australia, Queensland, Rockingham Bay, *Dallachy* s.n. (MEL, holo., K!, iso.)

Tree to 40 m with red-brown fibrous bark throughout. Leaves discolorous; juvenile ovate; adult lanceolate to broadly lanceolate, 10–15 cm long, 2–4* cm wide; petiole channelled, 1.5–2.5 cm long. Umbels ± 7-flowered (occasionally only 3); peduncle broadly flattened, 1–2.5 cm long; pedicels thick, angular, 1–9[12] mm long or rarely absent. Buds fusiform; operculum conical, rostrate, or hemispherical, 10–12 mm long, 6–10 mm wide. Calyx-tube obconical, 6–8 mm long, 6–10 mm wide. Fruits hemispherical or obconical, 0.7–1.6 cm long, 0.7–1.5 cm wide with prominent level disc and 4 strongly exserted valves [rarely just included] [fruits characteristic with an oblique rim and a flange 2.5–4 mm tall]. Fig. 9/29.

KENYA. Kiambu District: Muguga Forest Station, Hort. Pudden, 16 Dec. 1952, *Greenway* 8756!
TANZANIA. Lushoto District: Amani, Drackenberg Plantations 1, 29 July 1930, *Greenway* 2318!
DISTR. **K** 3, 4; **T** 3

NOTE. Material previously determined as *E. resinifera* Sm. var. *grandiflora* Benth. (and treated as a synonym of *E. resinifera*, T.T.C.L.: 317 (1949)) is undoubtedly *E. pellita* F. Muell. as cited in the same work on p. 372.

30. **E. pilularis** *Sm.* in Trans. Linn. Soc. London 3: 284 (1797); T.T.C.L.: 372 (1949); Fl. Austral. 19: 156, fig. 57/u–v (1988); Friis in Fl. Eth. 2 (2): 88, fig. 72.11 (1996). Type: Australia, New South Wales, Port Jackson, *White* s.n. (LINN, holo., BM!, NSW, iso.)

Tree to 70 m with grey-brown [red to black], fibrous bark on most of the trunk but white or yellow-grey and smooth above. Juvenile leaves opposite, sessile, broadly lanceolate to lanceolate; adult lanceolate, 9–16 cm long, 1.6–3 cm wide; petiole flattened or channelled, 1–2 cm long. Umbels 7–15-flowered; peduncle flattened, 1–1.7[2] cm long; pedicels angular, [2.5]3–6[7] mm long. Buds clavate or fusiform; operculum conical or rostrate, 4–5 mm long, 3–5 mm wide. Calyx-tube obconical, 3–4 mm long, 3–5 mm wide. Fruits hemispherical or subglobose, 6–11 mm long, 7–11 mm wide with narrow to wide, ascending to descending disc and 4 flush or less often included valves. Fig. 9/30.

KENYA. Uasin Gishu District: Kaptagat, 14 Nov. 1953, *Pudden* 17!
TANZANIA. Lushoto District: Lushoto Arboretum, 26 Jan. 1967, *Semsei* 4187!
DISTR. **K** 3–5; **T** 3

NOTE. Blackbush is one of the principal hardwoods of Australia yielding hard strong durable timber.

31. **E. polyanthemos** *Schauer* in Walp., Repert. Bot. Syst. 2: 924 (1843); T.T.C.L.: 371 (1949); Fl. Austral. 19: 413, fig. 107/a–b (1988). Type: Australia, New South Wales, N of Bathwest, *Cunningham* 136 (ubi?, holo., BM!, K!, iso.)

---

* Fruiting isotype has elliptic leaves up to 8.3 cm wide.

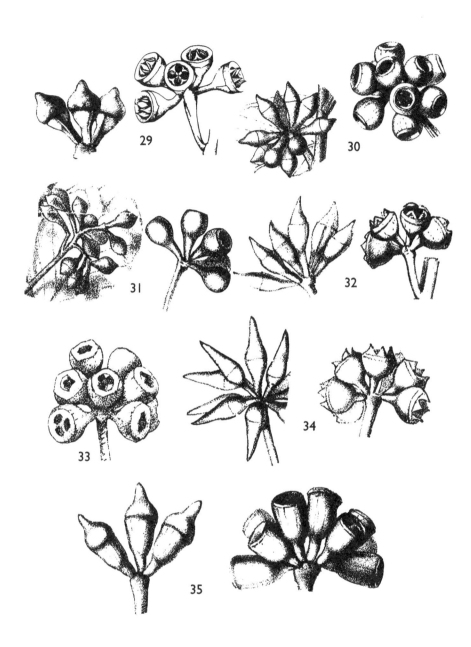

FIG. 9. Buds and fruits of *EUCALYPTUS* species, all natural size; species number as in main text.
29, *E. pellita;* **30**, *E. pilularis;* **31**, *E. polyanthemos;* **32**, *E. punctata;* **33**, *E. regnans;* **34**, *E. resinifera;* **35**, *E. robusta.* Illustrations reproduced courtesy of the former Australian Forestry & Timber Bureau, now CSIRO Forestry & Forest Products.

Tree to 25 m with grey-brown fibrous bark on trunk and larger branches or smooth bark throughout. Juvenile leaves round, emarginate; adult grey or glaucous, ovate to broadly lanceolate, 5.5–9 cm long, 1.5–3.5 cm wide* with intramarginal vein up to 4 mm from the margin; petiole terete or flattened, 1.5–2.5 cm long. Umbels 7-flowered, in terminal panicles; peduncle terete, 5–10 mm long, pedicels 2–5[7] mm long. Buds clavate to fusiform, glaucous; operculum conical, 1–2 mm long, ± 2 mm wide. Calyx-tube obovoid to obconical, ± 3 mm long and wide. Fruits obconical to pyriform, 4–7 mm long, 3–6 mm wide, often glaucous with broad descending disc and 3–4 included valves. Fig. 9/31, p. 45.

KENYA. Without locality, *Pudden* 27!
TANZANIA. Lushoto District: Amani, Drackenberg Plantation 1, 5 July 1930, *Greenway* 2314!
DISTR. **K** 3–5; **T** 3

NOTE. Has been misidentified as *E. camphora* in East Africa.

32. **E. punctata** *DC.*, Prodr. 3: 217 (1828); T.T.C.L.: 372 (1949); Dale, Introd. Trees Uganda: 38 (1953); Fl. Austral. 19: 204, fig. 65/k–l (1988). Type: Australia, 'New Holland', *Sieber* 623 (G, holo., G, NSW, iso.)

Tree to 35 m with cream to orange, later grey or grey-brown usually smooth matt bark. Juvenile leaves lanceolate; adult discolorous, lanceolate, slightly falcate, 8–15 cm long, 1.6–3[3.5] cm wide, shining above; petiole channelled or flattened, 1.5–2.6 mm long. Umbels 7-flowered; peduncle flattened or angular, 0.5–2 cm long; pedicels angular, 2–9[10] mm long. Buds ovoid to ± cylindrical; operculum conical or hemispherical or slightly rostrate, 5–6 mm long, 4–6 mm wide. Calyx-tube obconical or hemispherical, 4–5 mm long, 4–6 mm wide. Fruits hemispherical or cylindrical, 5–12 mm long, 5–10 mm wide with moderately broad level or ascending disc and 3–4 exserted valves. Fig. 9/32, p. 45.

UGANDA. Introduced into Ankole (1930–31) fide Dale
KENYA. Kiambu District: Muguga Arboretum, 28 June 1963, *Verdcourt* 3669G!
TANZANIA. Arusha District: Olmotonyi Forestry School, Feb. 1979, *Lema* 47!
DISTR. **U** 2; **K** 3–5; **T** 2, 3

33. **E. regnans** *F. Muell.* in Ann. Rep. Victorian Acclim. Soc.: 20, in obs. (1870–71); Fl. Austral. 19: 158, fig. 58/c–d (1988); Friis in Fl. Eth. 2 (2): 88, fig. 72.12 (1996). Type: Australia, Victoria, Dandenong, *Boyle* s.n. (MEL, lecto., fide Willis in Muellera 1: 167–168 (1967))

Tree to 75(–100) m with rough fibrous bark to about halfway up the trunk, then with white or grey-green smooth bark above. Juvenile leaves ovate to broadly lanceolate, oblique; adult lanceolate, 9–14 cm long, 1.6–2.7 cm wide with nervation very oblique; petiole channelled, 1.2–2.2 cm long. Umbels paired in leaf axils, 9–15-flowered; peduncle angular, 0.5–1.3 cm long; pedicels 2–4[5] mm long. Buds clavate; operculum conical, 2–3 mm long, 3–4 mm wide. Calyx-tube obconical, ± 3 mm long, 3–4 mm wide. Fruits obconical to pyriform, 5–9 mm long, 4–7 mm wide, with broad level or slightly descending disc and 3 flush or slightly exserted valves. Fig. 9/33, p. 45.

KENYA. South Nyeri District: Ragati Forest Station, compt 1D, June 1948, *F.D. in Bally* 6242!
TANZANIA. Lushoto District: Lushoto Arboretum, plot 116, 27 Dec. 1966, *Semsei* 4165!
DISTR. **K** 3, 4; **T** 3

* The 6–10 × 5–12 cm of my 1963 key must presumably include measurements taken from juvenile leaves.

NOTE. Reported to be the tallest hardwood in the world and the tallest tree in Australia; a record from the 19th Century of 132.5 m would, if correct, be the tallest tree ever recorded.

34. **E. resinifera** *Sm.*, in J. White, John White's Voyage: 231 (1790); T.T.C.L.: 372 (1949); Dale, Introd. Trees Uganda: 39 (1953); Fl. Austral. 19: 202, fig. 65/e–f (1988); Friis in Fl. Eth. 2 (2): 92, fig. 72.21 (1996). Type: Australia, New South Wales, Port Jackson, *J. White* s.n. (?ubi, holo., BM!, iso. fide Chippendale)

Tree to 45 m with red-brown fibrous bark throughout. Juvenile leaves ovate to broadly lanceolate; adult discolorous, broadly lanceolate, 10–17 cm long, 1.8–3.5 cm wide; petiole channelled, 1.5–3 cm long. Umbels 7–11-flowered; peduncle flattened, 1–2.2 cm long; pedicels angular, 3–10 mm long. Buds rostrate; operculum long and narrow, conical or rostrate, 1–1.2 cm long, 5–7 mm wide. Calyx-tube hemispherical, 4–5 mm long, 5–7 mm wide. Fruits hemispherical or obconical, [$^3/_4$-globose-turbinate] 6–11 mm long, [6]7–10 mm wide with level or obliquely convex disc and 4–5 exserted valves. Fig. 9/34, p. 45.

UGANDA. Ankole District: Ankole fide Dale
KENYA. Kisumu-Londiani District: Lumbwa Hill, *Nottidge* s.n. (fide EA)
TANZANIA. Lushoto District: Amani, Drackenberg Plantation 1, 25 July 1930, *Greenway* 2315!
DISTR. **U** 2; **K** 3–5; **T** 2, 3, 6, 7

35. **E. robusta** *Sm.*, Spec. Bot. New Holland: 39 (1795); T.T.C.L.: 373 (1949); U.O.P.Z.: 249, fig. (p. 250) (1949); Dale, Introd. Trees Uganda: 39 (1953); Amshoff in Fl. Gabon 11: 33 (1966); F. White in F.Z. 4: 209, t. 46/f (1978); Fl. Austral. 19: 200, fig. 64/k–l (1989); Friis in Fl. Eth. 2 (2): 92, fig. 72.6/11–12 (1996). Type: Australia, New South Wales, Port Jackson, *J. White* s.n. (LINN, holo., BM!, G, iso.)

Tree to 30 m with red-brown soft spongy subfibrous rough bark throughout. Juvenile leaves ovate; adult discolorous, broadly [ovate-lanceolate] lanceolate, 10–16[20] cm long, 2.7–4.5[8] cm wide; petiole 2–3.5 mm long. Umbels 9–15-flowered; peduncle broadly flattened, 1.3–3[3.5] cm long; pedicels angular, 1–9[12] mm long or absent. Buds rostrate or ± fusiform; operculum conical, rostrate, 10–12 mm long, 6–8 mm wide. Calyx-tube obconical, 6–7 mm long, 6–8 mm wide. Fruits cylindrical, [8]10–18 mm long, 6–11[12] mm wide, sometimes slightly constricted in the middle or near apex, with broad descending disc and 3–4 valves usually included and joined across the orifice or flush or slightly exserted. Fig. 9/35, p. 45.

UGANDA. Masaka District: Masaka, 5 May 1972, *Lye* 6835!
KENYA. Nakuru District: 6.5 km W of Njoro, 6 May 1975, *Nightingale* in *E.A.* 15867!
TANZANIA. Buha District: Kasulu, Feb. 1955, *Procter* 355!; U.O.P.Z.: 249 (1949) records this species from Migombani Gardens, Kizimbani and in the prison plantations in Zanzibar
DISTR. **U** 4; **K** 3–5; **T** 2–4, 6, 7; **Z**

NOTE. The Nightingale specimen is said to be "smooth ironbark tree" but buds, fruits and foliage are typical. *Katende & Lye* 306 from between Rushasha and Rwanyampazi, Ankole is probably a hybrid with *E. tereticornis*.

36. **E. rudis** *Endl.* in Endl. et al., Enum. Pl. Hügel: 49 (1837); T.T.C.L.: 372 (1949); Dale, Introd. Trees Uganda: 39 (1953); Fl. Austral. 19: 329, fig. 92/a–b (1988). Type: Western Australia, King George Sound, *von Hügel* s.n. (W, holo.)

Tree to 20 m with rough fibrous bark on trunk and lower branches, smooth above. Juvenile leaves ovate or round; adult lanceolate, 9–14 cm long, 1.3–3 cm wide; petiole terete or quadrangular, 1.3–3 cm long. Umbels 7-flowered; peduncle terete or quadrangular, 0.6–1.5[2] cm long; pedicels 4–8[13] mm long. Buds ovoid to fusiform; operculum conical, 6–8 mm wide. Fruits hemispherical, [globose] obconic or campanulate, 4–6 mm long, 6–9 mm wide with broad usually level or sometimes slightly convex or descending disc [often with flanged rim], valves exsert; seed brown to black, pitted. Fig. 10/36.

UGANDA. Ankole fide Dale
KENYA. Kisumu-Londiani District: Londiani compt Mt Blackett 3(b), 26 Nov. 1957, *Forestry Training School* 9!; Laikipia District: Rumuruti, 18 May 1953, *Smart* 119/93 (fide EA)
TANZANIA. Ngara District, 11 Nov. 1957, *Willan* 314!
DISTR. U 2; K 3–5; T 1–3

NOTE. Closely related to *E. camaldulensis* and according to Fl. Austral. shows a clinal change to that species in parts of Western Australia.

37. **E. saligna** *Sm.* in Trans. Linn. Soc. London 3: 285 (1797); Dale, Introd. Trees Uganda: 39 (1953); Jex-Blake, Gard. E. Afr. ed. 4: 243 (1957); Fl. Austral. 19: 198, fig. 64/g–h (1988); Friis in Fl. Eth. 2 (2): 91, fig. 72.20 (1996). Type: Australia, New South Wales, Port Jackson, *J. White* s.n. (LINN, holo., BM!, G, iso.)

Tree to 55 m with white or blue-grey smooth bark save for basal 4 m or so which has grey-brown rough flaky bark. Juvenile leaves ovate to broadly lanceolate; adult discolorous, lanceolate, 9–17 cm long, 2–3 cm wide; petiole channelled, 1.5–2.5 cm long. Umbels 7–11-flowered; peduncle flattened, 0.4–1.8[2.2] cm long; pedicels absent or angular, up to 3[8] mm long. Buds fusiform or ± ovoid; operculum conical, 3–4 mm long and wide. Calyx-tube hemispherical, cylindrical or campanulate, 2–3 mm long, 3–4 mm wide. Fruits cylindrical, campanulate or subpyriform, 5–8 mm long, 4–7[8] mm wide with narrow descending disc and 3–5 exserted [or included] erect valves. Fig. 10/37.

UGANDA. Mbale District: Sebei, N Elgon C. Forest Reserve, near Kapkweta Forest Station, 12 Jan. 1963, *Styles* 309!
KENYA. Central Kavirondo District: Maseno, 18 Jan. 1953, *Pudden* 6!
TANZANIA. Arusha District: Olmotonyi Forestry School, Feb. 1979, *Lema* 52!
DISTR. U 3; K 4, 5; T 2, 3

NOTE. A possible hybrid with *E. tereticornis* has been collected in Muheza District and hybrids with *E. grandis* are recorded from Uganda and Kenya as well as many just stated to be '*saligna* hybrids'.

38. **E. siderophloia** *Benth.*, Fl. Austral. 3: 220 (1867); Dale, Introd. Trees Uganda: 40 (1953); Fl. Austral. 19: 405, fig. 104/m–n (1988). Type: Australia, Queensland, Moreton Bay, *Cunningham* 51 (K!, lecto., BM!, FRI, NSW, W, isolecto., chosen by Johnson in Contr. New S. Wales Herb. 3: 117 (1962))

Tree to 45 m with rough grey to grey-black ironbark on trunk and at least larger branches [hard, deeply furrowed] but sometimes smooth above. Juvenile leaves ovate to broadly lanceolate; adult lanceolate, 8–15 cm long, 1–2 cm wide; petiole terete, 1–1.7 cm long. Umbels 7-flowered in terminal panicles; peduncle quadrangular, 6–12[23] mm long; pedicels [0]2–4 mm long. Buds fusiform; operculum conical, 3–4 mm long and wide, sometimes rostrate. Calyx-tube obconic, 2–4 mm long, 3–4 mm wide. Fruits obconical, [4]6–8 mm long, [4.5]5–7[7.5] mm wide with narrow descending disc and usually 4 exserted valves [usually rugose and ribbed]. Fig. 10/38.

UGANDA. Ankole District: Ankole 1930–31 fide Dale
KENYA. Widely recorded by Verdcourt (1963)

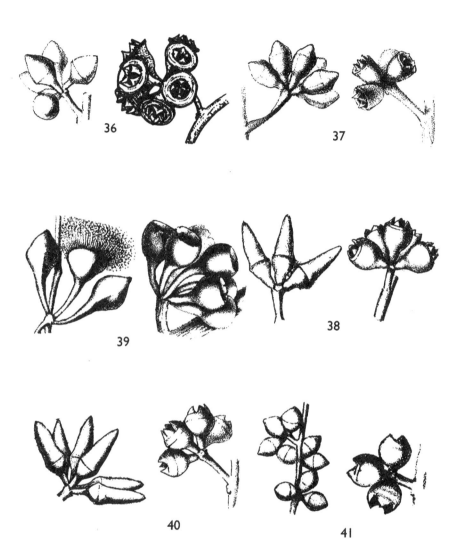

FIG. 10. Buds and fruits of *EUCALYPTUS* species, all natural size; species number as in main text. **36**, *E. rudis*; **37**, *E. saligna*; **38**, *E. siderophloia*; **39**, *E. sideroxylon*; **40**, *E. tereticornis*; **41**, *E. viminalis*. Illustrations reproduced courtesy of the former Australian Forestry & Timber Bureau, now CSIRO Forestry & Forest Products.

TANZANIA. Doubtfully recorded from Lushoto District: Amani
DISTR. **U** 2; **K** 3–5; **T** 3?

NOTE. No East African material at Kew or EA.

39. **E. sideroxylon** *Woolls* in Proc. Linn. Soc. New S. Wales 11: 859 (1887); Dale,
Introd. Trees Uganda: 40 (1953); Fl. Austral. 19: 425, fig. 109/c–d (1988); Friis in Fl.
Eth. 2 (2): 106, fig. 72.53 (1996). Type: Australia, New South Wales, Lachlan R.,
Cunningham 205 (K!, holo., BM!, iso.)

Tree to 35 m with black [jet black or very dark red-black] hard deeply furrowed
bark throughout or on trunk and larger branches and white smooth bark on upper
branches. Juvenile leaves linear to lanceolate (or ovate); adult grey, lanceolate, 7–14
cm long, 1.2–1.8 cm wide; petiole terete, 1–2 cm long. Umbels [3]7-flowered;
peduncle quadrangular to terete, [0.5]0.7–2 cm long, pedicels 0.2–1.5[2] cm long.
Operculum 3–5 mm long, 4–5 mm wide. Calyx-tube 4–6 mm long, 4–5 mm wide.
Fruits ovoid, subglobular or urceolate, 5–11 mm long, 6–10 mm wide with narrow
descending disc obscured by broad, caducous staminal ring, and 5 included valves.
Fig. 10/39, p. 49.

UGANDA. Ankole District: Ankole fide Dale
KENYA. Uasin Gishu District: Kitale, Nov. 1953, *Pudden* 33 & 34!
TANZANIA. Morogoro District: Uluguru Mts, Chigurufumi Forest Reserve, Mar. 1955, *Semsei*
    2008!
DISTR. **U** 2; **K** 3–5; **T** 3, 6

NOTE. Yields an extremely durable strong wood. Subsp. *tricarpa* L.A.S. Johnson has 3-flowered
    umbels; *Greenway* 8760, Kenya, Karura Forest Reserve compt 15, 424 (1909) and *Siemens* 26
    from **K** 3, Auea Spring (not traced) would appear to be this.

40. **E. tereticornis** *Sm.*, Spec. Bot. New Holland 1: 41 (1795); T.T.C.L.: 372 (1949);
Dale, Introd. Trees Uganda: 40 (1953); F. White in F.Z. 4: 210, t. 47/a (1978); Fl.
Austral. 19: 324, fig. 91/e–f (1988); Friis in Fl. Eth. 2 (2): 97, fig. 72.7/1–2 (1996).
Type: Australia, New South Wales, Port Jackson, *White* s.n. (LINN, holo., BM!, iso.)

Tree to 50 m with white, grey or grey-blue smooth bark throughout. Juvenile leaves
ovate; adult lanceolate, 10–20 cm long, 1–2.7 cm wide; petiole terete or channelled,
1.3–2.4 cm long. Umbels 7–11-flowered; peduncle terete or angular, 0.7–2.5 cm long;
pedicels [2]3–10 mm long. Buds conical; operculum conical, swollen at base, 8–13
mm long, 4–6 mm wide. Calyx-tube hemispherical, 2–3 mm long, 4–6 mm wide.
Fruits subglobular or ovoid, 5–7[8.5] mm long, 4–8[8.5] mm wide, with broad
steeply ascending disc and 4–5 strongly exserted valves; seeds black. Fig. 10/40, p. 49
& 11/5–6.

UGANDA. Busoga District: Mutai Plantation, compt 13, 5 June 1959, *Kingston* in F.D. 2202!
KENYA. Nairobi District: Karura Forest Reserve, compt 15A (1909), 14 Dec. 1952, *Greenway*
    8761!
TANZANIA. Arusha District: Olmotonyi Forestry School, Feb. 1979, *Lema* 54!
DISTR. **U** 1–3; **K** 1, 3–5; **T** 1–6; **Z**

SYN. *Leptospermum umbellatum* Gaertn., Fruct. & Sem. 1: 174, t. 35 (1788). Type: Queensland, Bay
    of Inlets, *Banks & Solander* (BM!, holo.)
    Eucalyptus umbellata (Gaertn.) Domin in Biblioth. Bot. 89: 467 (1928), *non* Dum.Cours.
    (1814)

NOTE. Yields hard strong durable wood. *Dale* U543 (Uganda, Ankole District, Ibanda, Mar.
    1948) is suggested by Ken Hill to be a hybrid between *E. tereticornis* and *E. dealbata* Schauer;
    Osmaston 2813 (Uganda, Toro District, Fort Portal) is the same.

FIG. 11. *EUCALYPTUS CAMALDULENSIS* — **1**, flowering branchlet, × ¹/₂; **2**, fruits, × 1;
EUCALYPTUS PANICULATA — **3**, flower-buds, × 1; **4**, fruit, × 1; *EUCALYPTUS
TERETICORNIS* — **5**, flower buds, × 1; **6**, fruits, × 1. 1, from *Cooling* 86; 2, from *White* 3488;
3, from *White* 6088; 4, from *Dillon* 8 MS; 5, from *Savory* 145; 6, from FHO 18537. Drawn by
J. Loken. From F.Z. 4, t. 47.

41. **E. viminalis** *Labill.*, New Holland Pl. 2: 12, t. 151 (1806); T.T.C.L.: 372 (1949); Fl. Austral. 19: 358, fig. 96/i–j (1988); Friis in Fl. Eth. 2 (2): 101, fig. 72.42 (1996). Type: Australia, Tasmania, probably Recherche Bay, *Labillardière* s.n. (FI, holo., L, iso.)

Tree to 50 m with grey, white or yellowish white smooth bark throughout or rough at base or on most of the trunk. Juvenile leaves opposite, sessile, lanceolate, cordate or amplexicaul; adult lanceolate, 12–20 cm long, 0.8–2 cm wide; petiole terete or slightly flattened, 1–2.5 cm long. Umbels 3- or 7-flowered; peduncle angular or flattened, 4–8[12] mm long; [pedicels 1–4 mm long]. Buds ovoid; operculum conical or hemispherical, 3–4 mm long, 3–5 mm wide, apiculate. Calyx-tube hemispherical or campanulate, 2–3 mm long, 3–5 mm wide. Fruits hemispherical to subglobular, 5–8 mm long, 5–9 mm wide with broad ascending disc and 3–4 exserted valves. Fig. 10/41, p. 49.

subsp. **cygnetensis** *Boomsma* in J. Adelaide Bot. Gard. 2: 295 (1980); Fl. Austral. 19: 359, fig. 96/k–l (1988). Type: S Australia, Cygnet R., Kangaroo I., *Hagerstrom & Boomsma* 510 (AD, holo.)

Rough bark more extensive; umbels 7-flowered.

KENYA. Recorded from Kiambu District: Muguga
TANZANIA. Lushoto District: Shagayu Forest Reserve, 7 Apr. 1964, *Willan* 633 & 634!
DISTR. **K** 3, 4; **T** 1, 3, 7

NOTE. *Muze* 54 from Lushoto Arboretum has been named *E. viminalis* var. *racemosa* F. Muell. but that has been raised to specific rank as *E. pryoriana* L.A.S. Johnson; it seems indistinguishable from typical *E. viminalis* apart from habit.

1. **PSIDIUM**

L., Sp. Pl.: 470 (1753) & Gen. Pl. ed. 5: 211 (1754)

Trees or shrubs. Leaves opposite (at least in ours). Inflorescence axillary, 1–3-flowered. Calyx unlobed and almost or completely concealing the corolla before anthesis, subsequently splitting irregularly into 4–5 lobes, persistent. Petals 4–5. Stamens numerous, free. Ovary imperfectly (2–)4–5(–7)-locular with 4–5 intrusive parietal placentas; ovules numerous. Fruit a berry with numerous angular seeds.

Over 100 species, mostly in tropical America but a few in the Pacific. Several are widely cultivated for their fruit; 5 have been seen from the Flora area, one the common guava very commonly planted and now naturalised in places, and the strawberry guava also apparently regenerating.

1. Leaves pubescent beneath; buds constricted at top of
   calyx-tube · 2
   Leaves glabrous or almost so; buds not or scarcely
   constricted · 3
2. Lateral nerves 12–20 pairs, prominent beneath and well
   differentiated from the smaller intermediate nerves;
   young growth densely pubescent; branches often 4-
   angled; buds completely closed at apex · · · · · · · · · · ·   1. *P. guajava*
   Lateral nerves 10 or fewer pairs or, if apparently more, the
   main and intermediate nerves not differentiated;
   branches terete; buds mostly with small aperture at the
   apex ·   *P. guineense*

3. Leaves narrowly to broadly obovate, widest above the middle; branchlets ± round or compressed; buds open at the apex; peduncle and margins of base of leaf with papilla-like hairs ·   2. *P. cattleianum*

   Leaves elliptic; branchlets 4-angled; buds closed, apiculate at apex ·4

4. Leaves obtuse to acuminate at apex; buds closed at apex, the calyx-limb becoming split into 2–3 lobes; fruit green   *P. montanum*

   Leaves distinctly acuminate; buds closed at apex, the calyx-limb circumscissile at opening remaining attached to tube on one side; fruit sulphur yellow · · · · · · · · · ·   *P. friedrichsthalianum*

P. friedrichsthalianum (*O. Berg*) *Nied.*, the Costa Rican guava, a species from S Mexico to Columbia, has been grown at Amani Nursery (T.T.C.L.: 378 (1949)) (Tanzania. Lushoto District: Amani, 4 Feb. 1939, *Greenway* 5857). A tree 6–12 m tall with quadrangular branches. Leaves elliptic or oblong-elliptic, 3.8–12 cm long, 2.5–5 cm wide, acuminate at the apex, cuneate at the base, glabrous (or puberulous?) beneath; flowers white, 2.5 cm wide, solitary on slender peduncles; fruit sulphur yellow, round or ellipsoid, 3–6 cm long.

P. guineense *Sw.*, the Brazilian guava, has been grown at Amani (T.T.C.L.: 379 (1949)) (Tanzania. Lushoto District: Amani, Kiumba Plantation, 22 Jan. 1931, *Greenway* 2842 and Mangula, near Pentecostal Church, 7 Apr. 1982, *Kisena & Mtui* 11). Shrub or tree 1–7 m tall, very similar to the common guava but with leaves 3.5–14 cm long, 2.5–8 cm wide with fewer lateral nerves; fruit round or pear-shaped, 1–2.5 cm wide.

P. montanum *Sw.*, a Jamaican guava (T.T.C.L.: 379 (1949)), has been cultivated at Amani. Tree 9–15(–30) m tall with elliptic, oblong-elliptic or elliptic-lanceolate leaves, 3–9.5 cm long, 1–3.5(–4) cm wide, obtuse to acuminate at the apex, cuneate to rounded at the base; flowers solitary but can appear to be in false racemes due to falling or non-development of leaves, smelling of bitter almonds; fruits green, subglobose, ± 2 cm in diameter, very fragrant.

Jex-Blake (Gard. E. Afr. ed. 4: 344 (1957)) mentions a '*P. araca*' and a '*P. coriaceum*' but I have seen no material.

1. **P. guajava** *L.*, Sp. Pl. 1: 470 (1753); Popenoe, Man. Trop. Fruits: 272, fig. 35 (1920); U.O.P.Z.: 423, fig. (p. 425) (1949); T.T.C.L.: 379 (1949); Dale, Introd. Trees Uganda: 62 (1953); Jex-Blake, Gard. E. Afr. ed. 4: 300 (1957); F.F.N.R.: 303 (1962); Amshoff in Fl. Gabon 11: 6 (1966); Boutique, F.C.B., Myrtaceae: 31 (1968); Amshoff in C.F.A. 4: 94 (1970); F. White in F.Z. 4: 185 (1978); Morton, Fruits Warm Climates: 356, t. L (1987); Thulin & G. Moggi in Fl. Somalia 1: 244 (1993); Friis in Fl. Eth. 2 (2): 72, fig. 72.2 (1996). Type: "India"* ; Herb. Clifford: 184, Psidium No. 1 (BM!, lecto., chosen by McVaugh, 1989)

Small tree to 12 m tall; bark pale brown or copper-coloured, thin, smooth, flaking off to reveal greenish layer beneath. Leaves elliptic to oblong-elliptic, 7–15 cm long, 3–5(–7) cm wide, rounded or acute at the apex, rounded at the base, densely pubescent beneath; lateral nerves 12–20 pairs, parallel, prominent beneath. Flowers axillary, solitary or in small clusters; calyx completely enclosing the young bud; lobes 9 mm long, 5 mm wide, reflexed in open flower, whitish tomentellous inside. Petals white, oblong-elliptic, 1.3 cm long, 8 mm wide. Stamens ± 250, ± 1 cm long. Style ± 1 cm long, slightly capitate at the apex. Fruit light yellow with a pink or crimson blush, globose, ovoid or pyriform, 5–10 cm long with granular flesh, juicy central white or red pulp and numerous very hard seeds 3 mm long.

---

* Although India is given as locality in Species Plantarum, 'Jamaica, Mexico, Brasil etc.' are given in Hortus Cliffortianus: 184 (1737).

FIG. 12. *PSIDIUM CATTLEIANUM* — **1**, habit, × ¹/₃; **2**, flower, × 2; **3**, fruit, × ¹/₂; **4**, fruit, cross-section, × ¹/₂. Drawn by J. Lindley. From R.B.G. Kew illustrations collection.

UGANDA. Mengo District: Kyempisi, Misindye, *Tanner* 6034!

KENYA. Machakos District: Kilungu Location, Kithembe Hill, Kalondu Farm, 14 Mar. 1982, *Mwangangi* 2234!

TANZANIA. Lushoto District: Amani, Plantation 2, 1 July 1929, *Greenway* 1632!, 1633! & Kungumi, 27 Mar. 1971, *Shabani* 674! & Chakechake Village, 14 Apr. 1969, *Ngoundai* 283!; Buha District: Kasulu, 20 Oct. 1954, *Kasele* 6014!

DISTR. **U** 4; **K** 4; **T** 3, 4; widely grown throughout East Africa and has become wild in various places in Tanzania especially down the coast and near habitations. Originally a native of South America, now widely planted throughout the tropics and subtropics

HAB. Coastal and near habitations inland; 1–1200 m

SYN. *P. pyriferum* L., Sp. Pl. ed. 2: 672 (1762). Type as for *P. guajava*

NOTE. Eaten raw and cooked, also made into jam and jelly etc. There are numerous cultivars.

2. **P. cattleianum** *Sabine* in Trans. Roy. Hort. Soc. 4: 317, t. 11 (1821); Popenoe, Man. Trop. Fruits: 279, fig. 36 (1920); A. Schroed. in Journ. Arn. Arb. 27: 314 (1946); U.O.P.Z.: 423 (1949); T.T.C.L.: 378 (1949); Dale, Introd. Trees Uganda: 61 (1953); Jex-Blake, Gard. E. Afr. ed. 4: 301 (1957); F.F.N.R.: 303 (1962); Amshoff in Fl. Gabon 11: 5 (1966); F. White in F.Z. 4: 186 (1978); Morton, Fruits Warm Climates: 363, fig. 99 (1987). Type: plant grown at Messrs. Barr & Brookes, Ball's Pond, Newington from seed received from China (ubi?)

Evergreen shrub or small tree 2–8 m tall with round branchlets. Leaves narrowly to broadly obovate-elliptic, 3.4–12 cm long, 1.6–6 cm wide, rounded to shortly acuminate at the apex, cuneate at the base, glabrous save for short papilla-like hairs on the petiole and margins of base of lamina; lateral nerves in ± 9 pairs. Flowers axillary, solitary or in pairs; buds slightly open at the apex, the calyx not quite enclosing the bud; lobes semicircular, 2 mm long, 3 mm wide, glabrous inside. Petals white, ± round, 5 mm diameter. Stamens 5 mm long. Style 5 mm long, distinctly capitate at the apex. Fruit dark purple-red (or yellow in var. *lucidum* Hort.), globose or obovoid, 2.5–4 cm long. Fig. 12.

UGANDA. 'Not uncommon in Uganda' (*Dale*)

KENYA. Nairobi Arboretum, Block 10, 8 Feb. 1952, *Dyson* 195!

TANZANIA. Lushoto District: Amani, 29 Oct. 1928, *Greenway* 948! & Drackenberg Plantation 1, 25 July 1930, *Greenway* 2308! & W Usambaras, Kitivo N Forest Reserve, 27 Mar. 1971, *Shabani* 697! & Lushoto, Jägertal, 10 May 1970, *Shabani* 550!

DISTR. Widely cultivated in East Africa. The strawberry guava, originally from South America, now widely cultivated throughout the tropics and subtropics

HAB. Cultivated and regenerating freely, also spread by birds (fide Greenway); 0–2100 m

NOTE. Used in the same way as the last species.

2. EUGENIA

L., Sp. Pl.: 470 (1753) & Gen. Pl., ed. 5: 211 (1754); Verdc. in K.B. 54: 41–62 (1999)

Trees, shrubs or sometimes small subshrubs from a woody rootstock; sometimes dioecious. Leaves opposite (at least in ours). Flowers axillary, hermaphrodite or functionally unisexual or purely male, solitary or fasciculate or in few-flowered cymes or racemes; bracteoles usually persistent; receptacle (calyx-tube) sharply differentiated from the pedicel, adnate to the ovary in hermaphrodite flowers, concave in male. Calyx-lobes 4–5, usually ciliolate, persistent. Petals 4–5, free, often ciliolate, sometimes persistent. Stamens numerous, included on the receptacle rim; filaments free. Ovary 2-locular with axile placentation; ovules numerous. Style as long as or longer than stamens in hermaphrodite or female flowers, with capitate or shortly 2-lobed stigma but often rudimentary or completely lacking in male flowers. Fruit a 1–3-seeded berry; cotyledons partly or completely fused.

A large genus "with several hundred species" (F.Z.) to "c. 1000" (Mabberley, The Plant Book, ed. 2 (1997)). Mostly in tropical and subtropical America but many also in Africa and Asia.

Several introduced species of *Eugenia* have been cultivated in East Africa and are briefly mentioned here and included in the key. Several other cultivated species formerly placed in *Eugenia* are now placed in *Syzygium, Myrciaria* etc.

1. Wild species* · 2
    Cultivated species · 12
2. Pyrophytic subshrub 7.5–50 cm tall from horizontal
    rhizome; leaves narrowly elliptic, lanceolate or
    oblanceolate, 4–11 × 1.1–3.5 cm (**T** 4, 7, 8) · · · ·   11. *E. malangensis*
    Forest shrubs or trees, often riverine · 3
3. Leaves sessile; branches, leaves and pedicels said to
    be glabrous; shrub to 5 m (**T** 3, E Usambaras) [no
    material seen; leaves said to be 'rather big' but no
    dimensions given and inflorescences racemose
    not fasciculate] ·   7. *E. scheffleri*
    Leaves with distinct but often short petiole · 4
4. Leaves ± cordate at the base (at least in Flora area) · · · · · · · · · · · · · · · · · 5
    Leaves rounded to cuneate at the base · 7
5. Leaves 14 × 7 cm; E Usambaras (3000 m) · · · · · · ·   see note after *E. scheffleri*
    Leaves small; not from E Usambaras · 6
6. Leaves ovate, broadly elliptic or almost rounded,
    0.9–4.7 × 0.4–3.5 cm, rounded at the apex; plant
    glabrous or slightly pubescent (**K** 7) · · · · · · · · ·   8. *E. nigerina*
    Leaves oblong, up to 9.2 × 4 cm, narrowed to a
    rounded apex; young stems, petioles, midrib
    beneath, pedicels and particularly calyx-tube
    densely shortly ferruginous hairy or velvety (**K** 4)   10. *E. thikaensis*
7. Pedicels with dense spreading hairs or short curled
    hairs (very rarely glabrous); leaves distinctly
    discolorous; swamp-forest etc. species from **U** 2–4;
    T 1–4 · 8
    Pedicels glabrous or if densely pubescent then an
    upland forest species 1700–2000 m from **T** 7 · 9
8. Pedicels with dense spreading white and
    ferruginous hairs 0.5–0.7 mm long; leaves 1–5 ×
    0.8–2.4 cm; branchlets densely ferruginous
    pubescent when young; midrib persistently
    pubescent beneath (**T** 4, Ugalla R.) · · · · · · · · ·   3. *E. aschersoniana*
    Pedicels very finely pubescent with ± curled very
    short hairs or rarely ± glabrous; leaves 3.4–9 ×
    1.3–5 cm; branchlets puberulous when young;
    leaves glabrous (**U** 2–4; **T** 1, 4) · · · · · · · · · · · · ·   2. *E. bukobensis*
9. Subrheophytic glabrous shrub 1–4 m tall with
    narrowly oblong-elliptic leaves 2.5–6.3 × 0.7–1.9
    cm (**K** 4, 7; **T** 6) ·   9. *E. tanaensis*
    Not as above or, if (rarely) a plant of rocky river
    beds etc., then leaves different · 10

* An imperfectly known species (nr. 4) from Kenya, Nzaui Mt, is omitted.

10. Lowland or coastal species in forest, woodland or
on dunes, coral etc. (0–310 m, rarely to 1000 m)
[leaves nearly always rounded] · · · · · · · · · · · ·   1. *E. capensis*
                                          subsp. *multiflora*

     Montane forest species (1250–2200 m in **T** 3 (W
Usambaras), **T** 6 (Ulanga) and **T** 7 (Mufindi,
Highlands etc.) · 11
11. Leaves small, elliptic to narrowly elliptic, 0.7–3.3 ×
0.5–2.3 cm, subacute with rounded tip; probably
always dioecious, the male flowers in 1–5-flowered
fascicles, the female usually solitary; peduncle
absent; pedicels glabrous or densely pubescent;
stems densely shortly ferruginous pubescent; fruit
ellipsoid, 9 × 7 mm (**T** 7, Mufindi Highlands;
1700–2000 m) ·   6. *E. mufindiensis*
     Leaves larger, 2.5–11.5(–13) × 1.5–5.5 cm, sharply
acuminate (in typical material from **T** 6, 7)
rounded, acute or narrowed to a rounded apex
(in Usambara material); pedicels and stems
glabrous (**T** 3, 6, 7; 1250–2200 m) · · · · · · · · ·   5. *E. toxanatolica*
12. Fruit with 7–8 fat ribs; leaves 2–6.3 × 1.3–3.3 cm,
acuminate; flowers solitary or in fascicles of ± 4 · ·   *E. uniflora*
     Fruit not ribbed · 13
13. Leaves 6.3–16 × 2.5–6.3 cm, obtuse at apex; flowers
solitary on pedicels 2.5–5 cm long; fruits 1.2–2 cm
diameter ·   *E. dombeyi*
     Leaves 3.3–6 × 1.3–2.3 cm, obtusely acuminate;
flowers in (1–)15–20-flowered fascicles on pedicels
4–5 mm long; fruits 3–4 mm in diameter · · · · ·   *E. capuli*

E. brasiliensis Lam. (= *E. dombeyi* (*Spreng.*) *Skeels*) known by a variety of similar names such as
'grubixameira', 'grumixama', 'grumichama' etc. was formerly cultivated in Tanzania at Mombo
(*Veith A.H.* 6678) and is recorded in gardening literature as being grown in Kenya. A slender
ornamental tree 7.5–15 m tall with reddish young shoots. Leaves ovate or oblong-obovate,
6.3–16 cm long, 2.5–6.3 cm wide, obtuse at the apex, cuneate at the base, coriaceous; petiole 1
cm long; flowers solitary; pedicels 2.5–5 cm long with 2 basal narrowly elliptic densely glandular-
pustulate bracts 1.3 cm long. Calyx-lobes green. Petals white, 1.5 cm long; fruit green turning
bright red to dark purple or nearly black, oblate, 1.2–2 cm diameter with persistent sepals 1.25
cm long, edible.

E. capuli (*Schlecht. & Cham.*) *O. Berg*, a native of Mexico and Guatemala has been cultivated
in several plantations at Amani (Tanzania, Lushoto District: *Greenway* 1602, 2955, *Soleman A.H.*
6131, *Ruffo* 643, 682 and *Furuya* 155; T.T.C.L.: 377 (1949)). A much-branched shrub 3–4.5 m
tall. Leaves elliptic or rhombic-elliptic, 3.3–6 cm long, 1.3–2.3 cm wide, narrowed to a narrowly
rounded tip or obtusely acuminate at the apex, cuneate at the base, densely glandular-
pustulate; petiole slender, 3–4 mm long; fascicles (1–)15(–20)-flowered; pedicels 4–5 mm long;
bracteoles at base of calyx 0.2 mm long. Calyx-lobes under 0.5 mm long. Petals white, elliptic, 2
mm long, 1.5 mm wide. Style long and slender, 4–4.5 mm long; disc 1 mm wide; male and
female flowers probably on separate plants. Fruit red turning black, subglobose, 3–4 mm in
diameter, edible but with very scanty pulp.

E. uniflora *L.* the Surinam or pitanga cherry (but also has many other names), a native of
Surinam, Guyana, French Guiana and S to Brazil and Uruguay has been cultivated at Entebbe
Botanic Gardens (Uganda, Mengo District: *Snowden* 1829), Nairobi (Kenya, Nairobi District:
Mathenge 463, *Williams Sangai* 548, *Mwangangi* 2027), Amani and Dar es Salaam (Tanzania,
Lushoto District: *Greenway* 934, 2922, *Ruffo* 626 & Uzaramo District: *Ruffo* 595), often as a hedge
plant (T.T.C.L.: 377 (1949)). Much-branched shrub or small tree 3–10 m tall, leaves ovate, 2–6.3
cm long, 1.3–3.3 cm wide, obtuse or obtusely acuminate at the apex, rounded at the base,
glabrous; petiole 2 mm long; flowers solitary or ± 4 in fascicles; pedicels slender, 1.5–2.5 cm long,

glabrous, with oblong brown gland-dotted bracts 7 mm long at base. Calyx-lobes 4 mm long, 1.5–2.5 mm wide, reflexed. Petals white. Stamens 50–60. Fruit green turning orange, bright red, deep scarlet or dark purplish maroon (almost black), oblate, 1.4–2.5 cm diameter, 7–8-ribbed, juicy and edible with either 1 large or 2–3 smaller seeds.

1. **E. capensis** (*Eckl. & Zeyh.*) *Sond.* in Fl. Cap. 2: 522 (1862); Sim, For. Fl. Port. E. Afr.: 68 (1909); Dummer in Gard. Chron. 52: 179 (1912); Palmer & Pitman, Trees S. Afr. 3: 1669, photos & fig. (p. 1670–1671) (1972); Ross, Fl. Natal: 260 (1972); F. White in F.Z. 4: 187, t. 42 pro parte (1978); K.T.S.L.: 125 (1994); A.E. van Wyk & P.S. van Wyk, Field Guide Trees S. Africa: 318, fig. 1 (p. 319) (1997); Verdc. in K.B. 54: 43 (1999). Type: South Africa, Bosjesmansrivier, Uitenhage, *Ecklon & Zeyher* 1772 (ubi?, holo., K!, iso.)

Several-stemmed shrub or small tree 1.5–7.5 m tall with pale grey-brown or brown fissured and flaking bark; wood hard; shoots glabrous. Leaves elliptic, narrowly elliptic or oblanceolate-obovate, 0.8–9.5(–13) cm long, 0.5–5.5(–7) cm wide, rounded or subacute or obtuse at the apex, rarely ± acuminate, cuneate to rounded-cuneate at the base, densely gland-dotted, very fragrant with a nutmeg smell when crushed, ± subcoriaceous, ± revolute; petiole 2–5 mm long. Flowers scented, in 1–20-flowered fascicles; pedicels (1–)3–13 mm long (can be short in both sexes); bracteoles usually triangular or ovate-elliptic, rarely lanceolate, up to 0.5(–1) mm long, glabrous. Calyx-tube 1–2.5 mm long, glabrous; lobes 4, rounded-elliptic or ± semicircular, unequal, 1.5–7 mm long, 1.5–6 mm wide, rounded at apex, gland-dotted. Petals white, elliptic or oblanceolate, 3–6 mm long, 2–3.5 mm wide, not ciliate or with very few ciliae. Male: stamens 30–55; filaments 2–3 mm long; anthers 0.7–1 mm long. Style remnant apparently absent. Disc ± shallowly 4-lobed, glabrous to pubescent. Hermaphrodite flowers similar: style 2.5–4 mm long; stigma peltate. Fruits red or purple to blue-black, subglobose, 5–9(–13) mm long, 5–8(–15) mm wide, pustulate, crowned by persistent calyx-lobes.

SYN. *Memecylon capense* Eckl. & Zeyh., Enum. Pl. Afr. Extratrop.: 274 (1836)

subsp. **multiflora** *Verdc.* in K.B. 54: 43, fig. 1 (1999). Type: Kenya, Kwale District, Jombo Mt, *Verdcourt* 5283 (K!, holo., EA!, iso.)

Leaves 2.5–13 cm long, 1.5–7 cm wide. Inflorescences usually (1–)3–20-flowered.

KENYA. Kwale District: Shimba Hills Lodge, Mkomba R., 23 Feb. 1988, *Robertson & Luke* 5165!; Kilifi District: Watamu, 3 May 1973, *Simpson* 30!; Lamu District: Boni Forest Reserve, 8.5 km E of Dodori R. at Magai, 30 Nov. 1988, *Robertson & Luke* 5640!
TANZANIA. Pangani District: Mafui, Mkwaja, 8 Oct. 1957, *Tanner* 3714!; Uzaramo District: Msasani, 23 Aug. 1939, *Vaughan* 2869!; Rufiji District: Mafia I., Kikuni, 14 Aug. 1937, *Greenway* 5092!; Zanzibar, Pwani, Mchangani, 1 Dec. 1930, *Greenway* 2610!
DISTR. **K** 7; **T** 3, 6, 8; **Z**; **P**; Mozambique
HAB. Coastal often ± dry evergreen forest and bushland, coastal thicket, mangrove swamp edges, flood plains, old termite mounds; 0–350(–1000) m

SYN. [*E. aschersoniana* sensu Engl. & Nied. in P.O.A. C: 287 (1895) & V.E. 3(2): 733 (1921) pro parte, *non* F. Hoffm. sensu stricto]
   [*E. capensis* (Eckl. & Zeyh.) Sond. subsp. *aschersoniana* sensu F. White in Kirkia 10: 403 (1977) & in F.Z. 4: 188 (1978) pro parte, *non* (F. Hoffm.) F. White loc. cit. sensu stricto; sensu Vollesen in Opera Bot. 59: 54 (1980); Thulin & G. Moggi in Fl. Somalia 1: 244 (1993), *non* (F. Hoffm.) F. White sensu stricto]
   E. taxon D, K.T.S.L.: 126 (1994) (material not seen)
   E. taxon F, K.T.S.L.: 126 (1994)

NOTE. Subsp. *capensis* occurs in South Africa and S Mozambique. Subsp. *multiflora* is very different from typical *E. capensis* from Cape Province which has small almost round leaves with revolute margins and invariably solitary flowers. Further up the coast and particularly on

Inhaca Island, what is clearly the same species displays much greater variability and larger leaves and 1–3-flowered fasciculate inflorescences are common. *E. mossambicensis* Engl. based on material from Beira is presumably similar although it has been kept up in some recent works on South African trees. Specimens occur in Mozambique which are very close to Tanzanian material, a fact noted by Greenway long ago (adnot. on *Greenway* 2610) so it is clear that the East African populations cannot be separated specifically from *E. capensis*; the problem that remains is how far subsp. *multiflora* extends southwards. The difficulty of placing intermediate collections does not in my opinion invalidate recognising subspecies for the extremes.

2. **E. bukobensis** *Engl.* in N.B.G.B. 2: 289 (1899); V.E. 3(2): 733 (1921); T.T.C.L.: 377 (1949); I.T.U. ed. 2: 272 (1952); F.F.N.R.: 302 (1962) pro parte; Boutique, F.C.B., Myrtaceae: 25 (1968); Verdc. in K.B. 54: 47 (1999). Types: Tanzania, Bukoba, *Stuhlmann* 3261, 3749, 3756, 3794, 3881 (B†, syn.)

Shrub or small tree 1.5–9 m tall; branchlets puberulous when young, purplish brown. Leaves usually drying distinctly discolorous, elliptic to elliptic-lanceolate, (2.5–)3.4–9 cm long, 1.3–5 cm wide, acute, subacuminate or even ± rounded at the apex, cuneate at the base, coriaceous, glabrous; petioles 3–4 mm long. Flowers in 2–8-flowered fascicles in axils of leaves or fallen leaves; pedicels 1.8–6.5 mm long, glabrous to puberulous with ± curled very short hairs; bracteoles placed immediately beneath the calyx-tube, ± triangular, 0.6 mm long, puberulous but not hidden by long hairs. Calyx-lobes ovate, 3.5 mm long, 3 mm wide, rounded, glabrous, with conspicuous gland-dots. Petals white, flushed pinkish mauve, pink, greenish yellow or green, oblong, obovate or ovate, elliptic, 4–7.5 mm long, 2–5.5(–6.5) mm wide. Stamens 35–40; filaments 2.5–7.5 mm long, longest in male flowers; anthers 0.5–1 mm long. Style 3–4 mm long, lacking in male flowers. Fruit at first green, then bright yellow becoming red-brown, ellipsoid, 1–1.2 cm long, 7.5–10 mm wide, minutely glandular, rugose but otherwise glabrous when adult but young fruits are usually ± densely puberulous with short adpressed curled hairs; seeds orange-brown, shining.

UGANDA. Ankole District: Ankole, 6 Sept. 1905, *Dawe* 375!; Kigezi District: Mitano Gorge, Nov. 1950, *Purseglove* 3500!; Mengo District: Entebbe, Buku, Nov. 1932, *Eggeling* 712!
KENYA. Central Kavirondo District: Central Nyanza, near Mundere, 29 Oct. 1956, *Trapnell* 2295
TANZANIA. Bukoba District: Nyakato, Apr. 1935, *Gillman* 260!; Biharamulo District: Mwerusi R., Nyakakiri, *Ford* 871!; Mwanza District: Kome I., Mar. 1924, *Bancroft* 186!; Buha District: Kasulu, near Heru Juu, Jan. 1955, *Procter* 352!
DISTR. U 2–4; K 5; T 1, 4; not known elsewhere (see note)
HAB. Swamp-forest, lakeside forest and derived thicket, often in rocky places, rock-cracks etc., also recorded from woodland on sandy loam; 1100–1800 m

SYN. [*E. cotinifolia* Jacq. var. *elliptica* sensu Engl. & Nied. in P.O.A. C: 287 (1895), *non* (Lam.) Engl. & Nied.]
  [*E. capensis* (Eckl. & Zeyh.) Sond. subsp. *nyassensis* sensu F. White; K.T.S.L.: 125 (1994), *non* (Engl.) F. White sensu stricto]

NOTE. Certainly some herbarium specimens have only female flowers but it is not certain whether the species can be both monoecious and dioecious. Some specimens do seem to have both male and female flowers but the matter should be investigated in the field. *Ford* 766 cited above has exceptionally large flowers noted by the collector as ³/₄" diameter. Miss Amshoff annotated it in 1957 "possibly a large-flowered form of *bukobensis*" which it undoubtedly is. F. White (F. Z. 4: 190 (1978)) tentatively sinks *E. bukobensis* into *E. capensis* (Eckl. & Zeyh.) Sond. subsp. *nyassensis* (Engl.) F. White but it is not the same as the isotype of *E. nyassensis* examined at Kew. I. Friis (Fl. Eth. 2(2): 75 (1996)) refers material from SW Ethiopia to *E. bukobensis*; it is certainly not identical with that from Bukoba but broadly conspecific and the same applies to material from Congo (Kinshasa) and Zambia, particularly around Lake Bangweolo. Whether it is possible to regard these populations as true subspecies needs further study but the differences are slight. All material I have seen from Congo (Kinshasa) and Ethiopia has glabrous pedicels. Nearly all from East Africa has pubescent pedicels. Ethiopian material can have considerably larger leaves.

There is a nomenclatural problem. When Engler first described *Eugenia bukobensis* in 1899 he cited in synonymy *E. cotinifolia* Jacq. var. *elliptica* (Lam.) Baker ex Engl. et Nied. in P.O.A. C: 287 (1895). Although this is no more than a reference to a misidentification it technically renders the name *E. bukobensis* illegitimate. Jacquin's species was established (Obs. Bot. 3:3 (1768)) for a species collected by *Gronovius* from an unknown locality and *E. elliptica* Lam., (Encyl. 3: 206 (1789)) was described from Mauritian material and bears no relationship to the African species. Conservation of the name *E. bukobensis* is being proposed.

3. **E. aschersoniana** *F. Hoffm.*, Beitr. Kennt. Fl. Zentr.-Ost-Afr.: 35 (1889); P.O.A. B: 221 (1895) pro parte; P.O.A. C: 287 (1895) pro parte; V.E. 3(2): 733 (1921); T.T.C.L.: 376 (1949); Verdc. in K.B. 54: 48 (1999). Type: Tanzania; Tabora District, Ugalla R., *Boehm* 75a (B†, holo.)

Shrub; branchlets densely ferruginous pubescent at first; older parts leafless, glabrous, with grey-brown slightly fissured bark. Leaves elliptic (or broadly obovate fide Hoffmann), 1–5 cm long, 0.8–2.4 cm wide, slightly acute to rounded at the apex, cuneate to rounded at the base, slightly pubescent when young and persistently so on the midrib beneath; petioles 3–5 mm long. Female flowers not seen; male flowers in 1–6-flowered fascicles in axils of leaves or fallen leaves; pedicels 0.6–1.4 cm long with dense spreading white and ferruginous hairs 0.5–0.7 mm long, slightly glandular; bracteoles minute, very densely ferruginous pubescent. Calyx-lobes 4, ovate, 3 mm long, 2.5–3 mm wide, ± rounded, pubescent outside, ciliate. Petals white, ovate or elliptic, 6–7 mm long, 4.8–5 mm wide, ciliate. Stamens ± 60 with filaments ± 3.5(–6) mm long. Style absent. Fruit not seen.

Tanzania. Tabora District: Ugalla R., Sept. 1881, *Boehm* 75a and Wala R., Ugalla R., July 1962, *Procter* 2091!
Distr. **T** 4; not known elsewhere
Hab. Riverine thicket; 1110 m

Syn. [*E. capensis* (Eckl. & Zeyh.) Sond. subsp. *aschersoniana* (F. Hoffm.) F. White in Kirkia 10: 403 (1977) & in F.Z. 4: 188 (1978), *non* sensu F. White]
[*E. capensis* sensu Ruffo et al., Cat. Lushoto Herb.: 222 (1996), *non* (Eckl. & Zeyh.) Sond.]

Note. F. White who saw no material of this taxon used the epithet for a very widespread subspecies of *E. capensis* which lacks the characteristic indumentum. Even if it is considered a subspecies the above taxon is a very different one. The Tabora area is known for many endemic taxa. The only specimen examined has male flowers.

4. **E.** sp.; Verdc. in K.B. 54: 49 (1999)

Shrub; young stems reddish, ridged, glabrous. Leaves discolorous, drying leaden-green above, elliptic or oblong-elliptic, 2.5–6.5 cm long, 1.3–3.5 cm wide, narrowed to a rounded tip, rounded to rounded-cuneate at the base, glabrous; petiole 3 mm long, strongly channelled above, not conspicuously gland-dotted. Inflorescences at least up to 4-flowered but flowers not known. Fruits drying blackish, globose, 9–12 mm diameter with calyx-lobes persistent, 3 mm long; pedicels 7 mm long, glabrous.

Kenya. Machakos District: crest of Nzaui Hill, 17 Mar. 1973, *Lawton* 1799!
Distr. **K** 4; not known elsewhere
Hab. Near remnant of *Newtonia* forest; the top of the hill is about 1800 m

Syn. *E. taxon A* of K.T.S.L.: 125 (1994)

Note. This somewhat resembles *E. natalitia* Sond. which F. White treats as a subspecies of *E. capensis* but is kept specifically distinct by all South African workers. Beentje *loc. cit.* states possibly equals *Luke* 393 from Embu District, Chogoria but I have not seen this specimen.

5. **E. toxanatolica** *Verdc.* in K.B. 54: 49, fig. 2 (1999). Type: Tanzania, Iringa District, Luheja Forest Reserve, 35°58'E, 8°21'S, *Fridmodt-Møller et al.* NG 075 (K!, holo., C!, iso.)

Small dioecious or ?monoecious tree 1.5–7.5 m tall or shrub to 3 m, with reddish or purplish brown stems; young shoots glabrous, with fissured epidermis red-brown beneath (or in one specimen from Kwiro minutely puberulous); slash brown, paler towards wood. Leaves only slightly discolorous, not or slightly coriaceous, elliptic, 2.5–11.5(–13) cm long, 1.5–5.5 cm wide, acute, ± obtuse, rounded or ± acuminate to sharply acuminate at the apex, cuneate at the base, glabrous, densely black gland-dotted beneath; petiole distinct, 5–8 mm long. Flowers sweetly scented, in 2–26-flowered fascicles; pedicels 0.5–1.8 cm long, gland-dotted, glabrous; bracteoles usually linear-lanceolate, 1.5 mm long, slightly pubescent or (in W Usambaras) minute or ± obsolete in some specimens. Calyx-tube narrow, 1.5–3 mm long; larger lobes elliptic, 2.2–6 mm long, 2.2–4 mm wide; smaller ± ovate, 1.5–3 mm long, 1.5–4 mm wide, all gland-dotted and not or scarcely ciliate. Petals white, elliptic, oblate or oblong, 3.5–6.5 mm long, 2.5–6 mm wide, reflexed. Disc densely shortly pubescent. Male flowers with ± 25–50 stamens with filaments 3–5 mm long; anthers 0.8 mm long. Style rudimentary, 0.5–2 mm long (or ± absent), pilose at the base. Female flowers with fewer shorter stamens. Style 4–5 mm long, much exceeding the filaments; stigma punctate or ± bifid-peltate. Fruit reddish, eventually almost black, ellipsoid or obpyriform, (0.6–)1.8 cm long, 7–13 mm wide, densely pustulate, crowned by persistent sepals; fruiting pedicels 0.8–1.6 cm long.

TANZANIA. Lushoto District: W Usambaras, Herkulu, 11 Feb. 1984, *Lovett* 241!; Mahenge District: Kwiro Forest Reserve, SW flank of ridge, 18 Jan. 1979, *Cribb et al.* 11013!; Iringa District: Mufindi, Lupeme Tea Estate, 31 Aug. 1971, *Perdue & Kibuwa* 11339!
DISTR. **T** 3, 6, 7; probably also in Malawi
HAB. *Podocarpus, Albizia* rain-forest and *Lasianthus, Polyscias, Craibia, Strombosia* forest in **T** 3; ridge-top moist montane forest in **T** 6 and **T** 7; 1250–2200 m

NOTE. I had at first considered that the W Usambara and southern populations could be separated as subspecies distinguished by long acutely acuminate leaves combined with rudimentary styles in male flowers up to 2 mm long contrasting with obtuse leaves and rudimentary styles ± absent. This does not work and I consider it is impossible to separate these populations. The W Usambara material has in the past been determined as *E. scheffleri* or *E. bukobensis* or left as an undescribed species. It will most probably be found in the Uluguru Mts.
   When this fascicle was in first proof P. Phillipson kindly read through it and loaned me eight fruiting specimens, one from the W Usambaras and seven from the Chome Forest in the S Pare Mts. *Phillipson* 5062 (W Usambaras, Mangamba to Gare, km 1) has fruiting inflorescences with 6–26 thin pedicels up to 1.8 cm long and more acuminate leaves, and is undoubtedly conspecific with the type from Mufindi. The material from the S Pare Mts has only 1–3 fruits per inflorescence on shorter thicker more corky pedicels and less acuminate leaves; it may represent a new species or subspecies, but I have referred rather similar specimens from throughout the range to *E. toxanatolica.* Until correlated fruit and flowers of both sexes are available from all the populations I do not propose to do more than mention the problem.

6. **E. mufindiensis** *Verdc.* in K.B. 54: 50, fig. 3 (1999). Type: Tanzania, Iringa District, Mufindi Highlands, Rufuna Forest Reserve, *Macha* 130 (K!, holo., NHT, iso.)

Shrub or small tree 1.8–3 m tall with reddish brown slightly flaky bark, probably always dioecious; stems often rather straggling, slender, reddish brown, densely shortly ferruginous-pubescent; internodes short; branchlets spreading. Leaves paler beneath, elliptic to narrowly elliptic, 0.7–3.3 cm long, 0.5–2.3 cm wide, subacute at apex but tip rounded, rounded at the base, glabrous save for few hairs on midrib beneath, densely gland-dotted beneath; petiole 1.5–2.5 mm long, slightly pubescent.

Flowers sweetly scented, female nearly always solitary, males in 1–5-flowered fascicles; peduncle absent; pedicels 0.3–1.2(–1.6) cm long, glabrous to densely pubescent; bracteoles 0.8 mm long. Calyx purplish green; tube narrow, 1–2 mm long; lobes 4, ovate or elliptic, 1.5 mm long, 1.2 mm wide, ciliate, densely covered with small raised orange-red opaque glands. Petals 4, white tinged pink or purplish, ± round or elliptic, 3 mm long, 2 mm wide, with similar glands to calyx. Stamens ± 38 in male flowers; filaments 1.2–2 mm long; anthers 0.6 mm long, the connective with a few glands. Style 4.8 mm long in female flowers but reduced to 2 vestiges 0.8 mm long in male flower. Disc ± glabrous. Fruit pink or red, ellipsoid, 9 mm long, 7 mm wide, crowned by disc and calyx-lobe bases. Fig. 13.

TANZANIA. Iringa District: Kigogo, 31 Oct. 1947, *Brenan & Greenway* 8245! & Luisenga Stream, 24 Aug. 1984, *D.W. Thomas* 3578! & Mufindi, 1 Feb. 1970, *Paget-Wilkes* 734!
DISTR. **T** 7; not known elsewhere
HAB. Montane forest of *Macaranga* etc., streamside forest of *Ilex, Trichocladus* etc. and forest edges; 1700–2000 m

NOTE. The distinctive nature of this species is emphasised by the fact that one specimen had been misdetermined as *Myrtus communis* L.! The sexuality of the inflorescences needs confirmation in the field.

7. **E. scheffleri** *Engl. & Brehmer* in E.J. 54: 331 (1917); V.E. 3(2): 732 (1921); T.T.C.L.: 377 (1949); Verdc. in K.B. 54: 52 (1999). Type: Tanzania, E Usambaras, Derema, *Scheffler* 213 (B†, holo.)

Shrub 4–5 m tall with stout glabrous greyish branches; internodes 5–6 cm long. Leaves oblong, rather large [no dimensions given], acute at the apex, subobtuse at the base, sessile; lateral nerves 9–10 on each side, joining not far from margin forming a submarginal vein, strongly prominent beneath but not above; venation remotely reticulate. Inflorescences axillary, racemose not fasciculate; flowers shortly pedicellate; pedicels glabrous, ± equalling flowers; bracteoles narrowly lanceolate, acute; receptacle obovoid, acute at the base, glabrous. Sepals broadly ovate, subacuminate, covered with glandular dots. Petals white, broadly elliptic, twice as long as sepals, ± rounded at the apex. Ovary 2-celled, each with 4–5 ovules; style robust, curved at apex; stigma small, disciform, oblique.

TANZANIA. Lushoto District: E Usambaras, Derema, Jan. 1900, *Scheffler* 213
DISTR. **T** 3; not known elsewhere
HAB. Dense shady forest on decomposing granitic soil with thick humus layer; 800–900 m

NOTE. This is known only from the original description (translated above) and one of E.G. Baker's very rough sketches (BM) which at least shows the leaves to be 14.5 × 6.8 cm and notes 'pedicels 4–6 mm'. The authors give the affinity to be with *E. kalbreyeri* Engl. & Brehmer from West Africa and Cameroon which has distinctly petiolate leaves but certainly pseudoracemose inflorescences. The name *E. scheffleri* has been applied to the species frequently in the W Usambaras and also tentatively to specimens from Iringa District but all these are at variance with the original description which mentions the 'absolute sitzenden Blätter'; this rules out any possibility of the name referring to these populations. Recently a specimen has been collected close to Derema which has subsessile, but subcordate leaves. Baker's sketch indicates a broadly rounded base; the inflorescence is unfortunately too young to be certain of its structure, but the flowers appear fasciculate with very short pedicels. I suspect this specimen - *Borhidi et al.* 87342, East Usambara Mts, 1 km below Derema Village in Hunga Valley, riverine forest, 800 m, 8 May 1987 - is a form of genuine *E. scheffleri*, confirming its distinctness. Derema is so close to Amani where quite intense collecting has been carried out for most of this century that it is curious the species has not been collected frequently; it is probably genuinely rare.

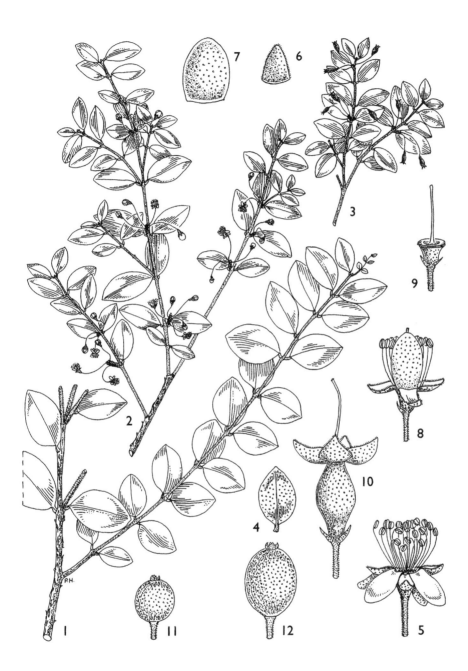

FIG. 13. *EUGENIA MUFINDIENSIS* — **1**, habit, × ²/₃; **2**, habit (male flowers), × ²/₃; **3**, habit (young fruit), × ²/₃; **4**, leaf (underside), × ²/₃; **5**, ♂ flower, × 4; **6**, calyx lobe, × 6; **7**, petal, × 6; **8**, ♀ flower (3 petals fallen), × 4; **9**, gynoecium, × 4; **10**, very young fruit, × 6; **11**, young fruit, × 2; **12**, fruit, × 2. 1, from *D.W. Thomas* 3578; 2, 5–7, from *Macha* 130; 3, 8–10, from *Paget-Wilkes* 734; 11, from *Lovett & Lovett* 2068; 12, from *Harris & Paget-Wilkes* 2401. Drawn by P. Halliday. From K.B. 54, p. 53.

8. **E. nigerina** A. Chev., Explor. Bot. Afr. Occ. Fr.: 268 (1920); Hutch. & Dalziel, F.W.T.A. ed. 1, 1: 199 (1927) & in K.B. 1928: 219 (1928); Aubrév., Fl. For. Soud.-Guin.: 85 (1950); Keay, F.W.T.A. ed. 2: 237 (1954); Roberty, Petite Flore de l'Ouest Africain: 247 (1954) [as "forme" of *E. coronata* Schum. & Thonn. but no combination made]; Verdc. in K.B. 54: 54 (1999). Type: Mali, between Dialiba and Sangana, *Chevalier* 260 (P!, lecto. (chosen here), K!, isolecto.)

Intricately branched shrublet 2–4(–8) m tall with reddish or brownish glabrous or shortly pubescent stems; bark grey, peeling. Leaves ovate, broadly elliptic or almost round, (0.9–)2–4.7 cm long, (0.4–)1.4–3.5 cm wide, rounded at the apex, rounded to slightly cordate at the base, glabrous, glandular-punctate, subsessile or petiole 1–1.5 mm long. Fascicles 1–6-flowered; bracteoles 0.5 mm long, glabrous or shortly pubescent; pedicels 0.8–1.4 cm long, glabrous. Sepals oblong, 2 large, 2 mm wide, 2 mm long, and 2 small, 1.5 mm wide, 1 mm long, rounded, glandular-punctate, glabrous. Petals white, 3–4 mm long. Rim of disc after filaments have fallen shortly pubescent. Style 3–3.5 mm long; stigma peltate, ± 1 mm wide. Fruits red, globose to ellipsoid, 1 cm long, 8 mm wide, glabrous, very densely glandular-pustulate.

Kenya. Kwale District: Between Samburu and Mackinnon Road, near Taru, 3 Sept. 1953, *Drummond & Hemsley* 4143!; Kilifi District: Ndzovuni R. gorge at ford 7 km N of Mtsengo, 23 Nov. 1989, *Robertson & Luke* 6083!; Tana River District: 1 km S of Bfunbe, 4 Aug. 1988, *Robertson & Luke* 5325!
Distr. **K** 7; Mali, Ivory Coast
Hab. Thicket in rocky places, river beds, gullies etc, *Spirostachys, Cynometra* flood plain thicket, forest, limestone caves with *Cynometra, Euphorbia, Lepisanthes* etc.; in West Africa Raynal states 'wooded grassland on alluvial plain and gallery forest 320 m' and Roberty 'in swamp submerged for 4–5 months'; 18–360 m

Syn. *E. taxon E*; K.T.S.L.: 126 (1994)

Note. Miss Amshoff identified *Drummond & Hemsley* 4143 as *E. nigerina* and I see no reason to doubt this on the evidence at present available. Several species (e.g. *Pseudovigna argentea* (Willd.) Verdc.) are widely disjunct, occurring in lowland East and West Africa and apparently not between. I have seen no fruiting material from West Africa. It has become usual to give the authority as *A. Chev. ex Hutch. & Dalz*iel but the original description appears to be adequate if somewhat sparse.

9. **E. tanaensis** Verdc. in K.B. 54: 55, fig. 4 (1999). Type: Kenya, Kitui District, Kindaruma Dam, *Bradshaw* 23 (K!, holo., EA!, iso.)

Dioecious shrub, probably predominantly rheophytic, 1–4 m tall, with glabrous stems; bark chestnut, ± flaky. Leaves tending to be held suberect, narrowly oblong-elliptic, 2.5–6.3 cm long, 0.7–1.9 cm wide, subacute but actual apex usually rounded, cuneate at the base, glabrous, gland-dotted; margin thickened and slightly revolute; petiole ± 3 mm long. Fascicles 1–5-flowered; pedicels ± 4 mm long, glabrous or pubescent; bracteoles triangular, 0.5 mm long, ferruginous pubescent. Sepals rounded oblong, 2–2.5 mm long, 1.8–2 mm wide, convex, ciliate, faces glabrous but with very conspicuous gland dots. Petals white or pinkish, oblong-elliptic, 3.5–5 mm long, 2–3 mm wide, glabrous except for a few apical ciliae. Stamens at least 40, filaments 2–2.5 mm long; anthers 0.8 mm long. Style 3 mm long, absent in male flower; stigma peltate. Fruits dark purple, 7–11 mm diameter, very closely glandular-pustulate, crowned by ± erect sepals; seeds solitary, subglobose–comma-shaped, 8 mm long, 7.5 mm wide.

Kenya. Kitui District: bank of Tana just below bridge on Kitui–Embu road, 27 Dec. 1970, *Gillett* 19276!; Kwale District: 5 km S of Mazeras, Mwachi, 10 Sept. 1953, *Drummond & Hemsley* 4260!; Kilifi District: Kombeni R. valley, edge of Kaya Fimboni, 21 Aug. 1989, *Robertson & Luke* 5813!
Tanzania. Kilosa District: at Great Ruaha R., 3 km S of junction with Yovi R., 7 Sept. 1970, *Thulin & Mhoro* 885!

DISTR. **K** 4, 7; **T** 6; not known elsewhere

HAB. On rocks at high water mark of river, rocky flood bed between low and high water marks; (5–)800–1000 m

SYN. *E. taxon B*; K.T.S.L.: 125, fig. (1994)
    E. taxon C; K.T.S.L.: 126 (1994)

NOTE. The lowland plant was found between large rocks in a river bed at only 5 m. The Tanzanian plant was on a sandy river bank at 450 m. The Kilifi District specimen is from mixed forest at 120 m and needs further investigation. Despite the difference in altitude the Kwale District material is the same species and from the same kind of habitat. The only difference noticed is that the sepals are not so conspicuously glandular-punctate. It is certain that the species is dioecious; *Bradshaw* 23 and *Thulin & Mhoro* 885 lack a style in the flowers examined. The style measurement is one from a fruit with a persistent style on *Gillett* 19276. One piece of *Thulin & Mhoro* 885 has puberulous pedicels but all the rest seen have them glabrous. This Tanzanian material has much more strongly gland-dotted leaves but without further collections the significance of this cannot be assessed.

10. **E. thikaensis** *Verdc.* in K.B. 54: 56, fig. 5 (1999). Type: Kenya, Thika, along the Chania R., below the Blue Posts Hotel, *Faden* 68/60 (K!, holo., EA!, iso.)

Shrub to 2.4 m; young stems (in part), petioles, underside of midrib, pedicels and particularly calyx-tube all densely short ferruginous-hairy or velvety; bark on twigs grey and fissured. Leaves oblong, 3–9.2 cm long, 2–4 cm wide, narrowed to a rounded apex, shallowly cordate at the base, with few hairs on midrib etc. above but soon glabrous, also a few hairs on blade beneath but only persistent on midrib; petioles 2 mm long. Fascicles 3–11-flowered; bracteoles 0.8 mm long, ciliate, but glabrous outside; pedicels 3–10 mm long, densely spreading pubescent; sepals 2 ± semicircular, 3.5 mm long, 2 mm wide, 2 ± circular, 4 mm long, 4 mm wide, all ciliate, concave and reflexed. Petals pinkish white, elliptic or oblong, 7–8 mm long, 4–5 mm wide, ciliate, reflexed at anthesis. Disc densely shortly pubescent. Stamens at least 50, variable in length in one flower, 2–6.5 mm long; anthers 1 mm long. Style 5 mm long, ± pilose; stigma 0.7 mm wide. Fruits unknown.

KENYA. Fort Hall District: Thika, along the Chania R., below the Blue Posts Hotel, 4 Apr. 1968, *Faden* 68/60! & 23 Mar. 1968, *Faden* 68/10! & Thika R. bridge N of Ol Doinyo Sabuk, about 5 km N of the Thika–Garissa road, N side of river, W of the bridge, 15 Nov. 1988, *Faden & Ng'weno* 88/12!

DISTR. **K** 4; not known elsewhere

HAB. Riparian forest of *Breonadia*, *Syzygium*, *Newtonia*, *Pachystela*, *Garcinia*, 'riverine tangle with patches of trees and shrubs'; 1400–1450 m

SYN. *E. taxon G*; K.T.S.L.: 126 (1994)

NOTE. The young leaves are wine-coloured and the stems reddish according to the collector. Purely male flowers have not been seen. A great deal of habitat of this species in the Thika area has been destroyed but hopefully it will be found elsewhere.

11. **E. malangensis** (*O. Hoffm.*) *Nied.* in E. & P. Pf. III (7): 81 (1893); Boutique, F.C.B., Myrtaceae: 26, t. 2 (1968); Amshoff in C.F.A. 4: 96 (1970); F. White in F.Z. 4: 190, t. 43 (1978); Troupin, Fl. Rwanda 2: 492, fig. 155/1 (1983) (as '*malanguensis*'); Verdc. in K.B. 54: 58 (1999). Type: Angola, Malange, *Mechow* 226 (B†, holo., BR, lecto., chosen by Amshoff)

Pyrophytic subshrub with caespitose shoots 7.5–50 cm tall (see note) from a horizontal rhizome, usually unbranched unless not burnt back, the stems ± densely short-pubescent. Leaves opposite, alternate, or in whorls of 3–4, ± shiny and darker above, narrowly elliptic, lanceolate, oblanceolate or linear, 4–11 cm long, 1.1–3.1 cm wide, obtuse at the apex, cuneate at the base. Flowers solitary in the axils or at base of plant sometimes in 2–9-flowered inflorescences, sweet-smelling; pedicels 0.6–2 cm

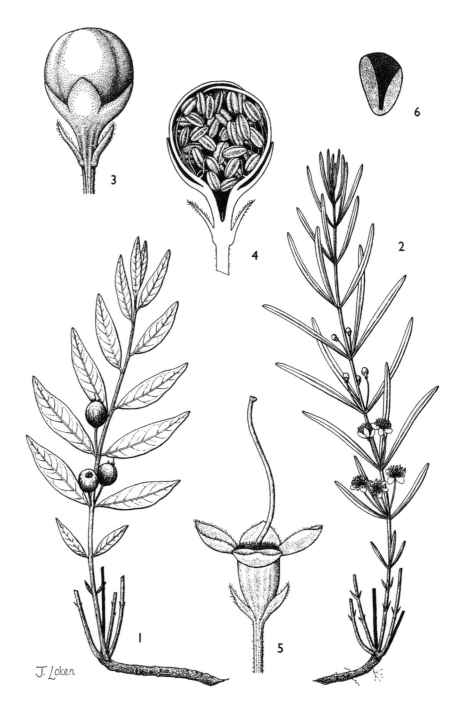

Fig. 14. *EUGENIA MALANGENSIS* — **1**, habit, broad-leaved variant, × $\frac{1}{2}$; **2**, habit, narrow-leaved variant, × $\frac{1}{2}$; **3**, flower bud, × 5; **4**, flower bud, median section, × 5; **5**, immature fruit, × 5; **6**, cotyledon, region of fusion lightly stippled, × 5. 1, from *Chapman* 263; 2, 5, from *Lawton* 1536; 3, from *Hoyle* 1211; 4, from *Norman* R4; 6, from *Brenan* 8027. Drawn by J. Loken. From F.Z. 4, t. 43.

long, pubescent; bracteoles lanceolate to triangular-ovate, 1–2 mm long, persistent. Calyx-tube obconic, 1.5–2 mm long; lobes obovate, round or rounded triangular, 1.5–3 mm diameter, ciliate, glabrous or with few hairs on the faces. Petals white or yellow, oblate-obovate or ± round, 4–7 mm long, 3–5 mm wide, ciliate or not. Stamens ± 50; filaments 3–5 mm long; anthers 0.8–1.2 mm long in male flowers, shorter in female. Receptacle glabrous between the stamens. Style 5 mm long in female flowers with capitate stigma 1 mm long, lacking or much reduced in male flowers. Fruits bright red, purple or black, ellipsoid or subglobose, 0.8–1.5 cm long, edible; pedicels 6 mm long, puberulous. Fig. 14.

TANZANIA. Buha District: 144 km from Kibondo on road to Kasulu, 15 July 1960, *Verdcourt* 2842!; Ufipa District: Pito, 25 Nov. 1949, *Bullock* 1940!; Iringa District: Mufindi, Mgololo area, 26 Nov. 1980, *Ruffo* 1620!; Songea District: Matengo Hills, Miyao, 6 Nov. 1956, *Semsei* 2572!
DISTR. T 4, 7, 8; Congo (Kinshasa), Rwanda, Burundi, Angola, Malawi, Mozambique and Zimbabwe
HAB. Open scrub, *Brachystegia* woodland, *Hyparrhenia* grassland subject to burning, rocky places in grassland, abandoned cultivations on black soil; 1350–2100 m

SYN. *Myrtopsis malangensis* O. Hoffm. in Linnaea 43: 134 (1881)
   Eugenia angolensis Engl. in N.B.G.B. 2: 288 (1899); Baker f. in J.L.S. 40: 71 (1911); R.E. Fr., Wiss. Ergebn. Schwed. Rhod.-Kongo-Exped. 1: 175 (1914); T.T.C.L.: 376 (1949); F.F.N.R.: 302 (1962). Type as for *E. coronata* var. *salicifolia*
   E. laurentii Engl. in N.B.G.B. 2: 288 (1899). Type: "Kongogebiet, *E. Laurent*" (?B†, holo.) (whether this is the same gathering as the *Laurent* from Lusambo-Lomami, 1895 cited by Boutique is not certain)
   E. marquesii Engl. in N.B.G.B. 2: 290 (1899); Boutique, F.C.B., Myrtaceae: 24 (1968); Amshoff in C.F.A. 4: 97 (1970). Type: Angola, Malange [Malandsche], *Marques* 343 (COI, holo., LISU, iso.)
   E. stolzii Engl. & Brehmer in E.J. 54: 330 (1917); Engl., V.E. 3 (2): 732 (1921); T.T.C.L.: 377 (1949). Types: Tanzania, Rungwe District, 'Konde Antuland' (fide Engler) [Kew sheets of 1798 have original label Ipyana (9°36'S, 33°52'E)] *Stolz* 1798 (B†, syn., K!, isosyn.), 1799 (B†, syn., BM!, K!, isosyn.)

NOTE. It is not certain quite how tall this plant grows. F. White suggested it never exceeded 50 cm and I think this is probably usually correct but Brenan (T.T.C.L.: 376 (1949)) under *E. angolensis* mentions 90 cm citing *Hornby* 64 as the source. Engler (V.E. 3(2): 732 (1921)) gives 6 m for *E. stolzii* and 2–6 m for *E. angolensis* but this could be an error for dm, a unit which he uses elsewhere on the page. Certainly no material reaching these heights has been seen.
   E. marquesii Engl. has been kept up by Boutique and Amshoff. It differs in having more than one flower in the leaf-axils and usually has short racemose inflorescences. F. White (F.Z. 4: 192 (1978)) points out that most specimens with such inflorescences appear to have escaped burning the previous season and occur sporadically throughout most of the range of *E. malangensis*; his decision to put it into synonymy is followed here but I have seen nothing matching *E. marquesii* from East Africa. Some plants undoubtedly have nothing but purely male flowers and the species may well be normally dioecious.

3. SYZYGIUM

Gaertn., Fruct. & Sem. 1: 166, t. 33 (1788), *nom. conserv.*

Mostly glabrous trees, shrubs or geoxylic subshrubs. Leaves opposite, penninerved, usually with an intramarginal vein. Inflorescences usually terminal and axillary. Flowers solitary and axillary or more usually in cymes or panicles, 4–5-merous; receptacle sometimes prolonged above the ovary and base usually narrowed to form a pseudopedicel above the bracteoles which are usually inconspicuous and deciduous. Calyx-lobes persistent or falling. Petals free or more or less united into a calyptra (in African species). Stamens usually numerous and conspicuous, in several series borne on receptacle rim, free or inconspicuously arranged in 4 obscure bundles; filaments filiform; anthers dehiscing by longitudinal lateral slits. Ovary 2 (rarely 3–4)-locular, the locules usually near distal part of ovary; ovules few to numerous in each locule, in crowded subcapitate clusters attached to the central partitions. Style filiform; stigma minute. Fruit a fleshy or dry leathery berry with 1–2 (rarely more) ± large seeds; testa membranous to crustaceous; cotyledons fleshy, usually completely free or partially fused (rarely completely fused).*

A large genus variously estimated at 500–800 species mostly in the Old World tropics but a few in Australia and New Zealand; several introduced and naturalised in America. Several species have been cultivated in East Africa, the most important being the clove tree *S. aromaticum* (L.) Merr. & Perry

There appears to be no single morphological character (or even series of characters) by which all species of the mostly New World group *Eugenia* can be conclusively separated from the exclusively Old World *Syzygium*. Evolutionary trends in the two groups, however, have proceeded in quite different directions; the two are readily recognisable by studies of ensembles of characters even though some of the very numerous species may be aberrant in some ways. *Eugenia* and *Syzygium* differ in at least 10 important morphological trends and the distinction between the two is supported by most recent studies of anatomy of wood, bark and flowers, vascular supply to the ovules; and by palynology.

1. Petals 1 cm long or more; cultivated or naturalised trees
    with large leaves, flowers and fruits · 2
   Petals smaller; wild, cultivated or naturalised · · · · · · · · · · · · · · · · · · · 4
2. Flowers in ± sessile clusters on new or old wood; petals
    1 cm long; leaves large, 16–34 × 5–14.5 cm; fruits
    2.5–6 cm long ·   *S. malaccense*
   Flowers in terminal and axillary cymes · 3
3. Leaves very acuminate at the apex, cuneate at the base,
    8–20 × 1.3–5.5 cm; petioles 5–10 mm long; petals 1.5
    cm long; fruit 2.5–5 × 2–4 cm (occasionally
    naturalised) ·   1. *S. jambos*
   Leaves obtuse, retuse or shortly acuminate at the apex,
    cuneate, rounded or slightly subcordate at the base;
    petioles 1–5 mm long; petals up to 1.2 cm long; fruits
    globose, 2 cm diameter · · · · · · · · · · · · · · · · · · ·   *S. aqueum*
4. Leaves cordate or subcordate at base and mostly quite
    rounded at the apex; petioles mostly short, often 2–5
    mm long · 5
   Leaves cuneate to rounded at the base, rounded to
    distinctly acuminate at the apex; petioles short or up
    to 4.5 cm long · 6
5. Leaves more oblong, often amplexicaul at base, rather
    thinner and more glaucous; inflorescences more
    open and calyx-tubes more obviously narrowed into
    pseudopedicels ·   4. *S. cordatum*
   Leaves more rounded, not amplexicaul, very
    coriaceous, not glaucous; inflorescences congested
    and calyx-tubes scarcely narrowed into
    pseudopedicels ·   9. *S. micklethwaitii*
                                                    (particularly subsp.
                                                subcordatum)

* Generic description and comments taken from unpublished 3rd volume of Hutchinson's Genera of Flowering Plants; this relied entirely on McVaugh for the Myrtaceae-Myrteae.

6. Leaves small, 1.5–4.5 × 0.9–2.4 cm, elliptic, narrowed to
     a rounded apex, cuneate at the base; petioles 2–5
     mm long; inflorescences 1.5–4 cm long; fruits
     subglobose, 0.9–1.2 × 0.9–1.3 cm (**T** 6, endemic to
     Uluguru Mts) ·   3. *S. parvulum*
   At least some leaves much larger, mostly with longer
     petioles and more extensive inflorescences · 7
7. Small pyrophytic subshrub 0.3–3 m tall; petioles usually
     short (1–)2–5(–9) mm long · · · · · · · · · · · · · · · ·   7d. *S. guineense*
                                                                    subsp. *huillense*
   Not as above · 8
8. Fire-resistant shrub or tree of ± dry woodland or
     grassland with scattered trees; leaves 6.5–14 × 3.6–7.7
     cm, often relatively wider than in other taxa with
     apex broadly rounded, subtruncate, emarginate or
     acute to shortly acuminate; petioles usually well
     developed, 1–2.5(–4.5) cm long · · · · · · · · · · · · · ·   7c. *S. guineense*
                                                                    subsp. *macrocarpum*
   Not as above · 9
9. Inflorescences few-flowered dichasia with total stalk
     (apparent pedicels) of flowers up to 1.7 cm long;
     true pedicels and inflorescence-axes very slender;
     leaves elliptic to oblanceolate, 2.5–6 × 1.3–2.7 cm,
     rounded to shortly acute at the apex, cuneate at the
     base; petals 4 mm long; fruits 1.3–1.5 × 0.8–1 cm;
     cultivated ·   *S. paniculatum*
   Inflorescences more congested with more numerous
     flowers and without other characters combined · · · · · · · · · · · · · · · 10
10. Leaves acute, acuminate or apiculate at the apex · · · · · · · · · · · · · · · 11
    Leaves rounded at the apex · 17
11. Leaves very narrowly attenuate into a long petiole, the
     stalk appearing up to 2.5 cm long but true petiole is
     thickened, up to about 8 mm long, and the rest
     exceedingly narrow leaf-base; buds crimson
     (cultivated crop tree) ·   *S. aromaticum*
    Leaf-bases not as above · 12
12. Calyx-lobes more developed, 1–2 × 2–3 mm, distinctly
     overlapping in bud and partly covering the calyptra
     of petals; leaves drying yellowish green, mostly
     obovate-oblong, rounded to a shortly acuminate
     apex (**U** 2, 4; **T** 1) ·   6. *S. congolense*
    Calyx-lobes very short and not overlapping, or if about
     as long then the leaves not drying yellowish green,
     mostly round or broadly elliptic · 13
13. Inflorescences mostly lateral in axils of fallen leaves on
     older stems (but some terminal present also); leaves
     mostly oblong-elliptic, 6–20 × 2.5–8.5 cm with main
     lateral nerves close and numerous; flowers drying
     pale orange-brown; fruit mostly oblong but
     sometimes ellipsoid or globose, the elongate ones
     sometimes slightly curved, 2.4–3.5(–5) cm long, 2
     cm wide (or ± 2 × 1 cm when dry) (widely planted
     tree, often naturalised in coastal areas) · · · · · · · · ·   2. *S. cumini*
    Inflorescences terminal and without other characters
     combined (native species) · 14

14. At least some leaves ovate, widest below middle and
    gradually narrowed to the apex; stilt roots at least
    sometimes present; swamp forest species, **U** 4
    (Masaka); **T** 7 (Sao Hill) · · · · · · · · · · · · · · · · ·   5. *S. owariense*
    Most leaves widest about the middle (but ovate leaves
    have been noted in a few specimens); stilt roots not
    recorded; riverine or evergreen forest species · · · · · · · · · · · · · · · 15
15. Leaves narrowly cuneate at the base, usually thinner;
    inflorescences more open with pseudopedicels more
    obvious and often quite slender · · · · · · · · · · · · · ·   7. *S. guineense*
    Leaves rounded at base, usually more coriaceous;
    inflorescences more congested with pseudopedicels
    less obvious or calyx-tube scarcely narrowed at base · · · · · · · · · · · · · · 16
16. Leaves very distinctly acuminate at apex; **T** 7 · · · · · · ·   8. *S. masukuense*
    Leaves slightly acuminate at apex (mostly rounded);
    K 7, **T** 3, 6 ·   9. *S. micklethwaitii*
                                                                         (and see 10. sp. A.)

17. Leaves more coriaceous, more rounded; petiole 1–5 mm
    long; inflorescences congested with calyx-tubes less
    narrowed into pseudopedicels; **K** 7 (Teita), **T** 3, 6 · · ·   9. *S. micklethwaitii*
    Leaves less coriaceous, more oblong or oblong-
    elliptic; petiole 2–6(–13) mm long; inflorescences
    more open with calyx-tubes distinctly narrowed
    into pseudopedicels · 18
18. Plant of **K** 7 (Shimba Hills); leaves subcordate to less
    often broadly cuneate with petioles 2–3(–5) mm long   4. *S. cordatum*
                                                           subsp. *shimbaense*

    Plants from throughout East Africa; leaves cuneate to
    rounded at base with petioles 2–6(–13) mm long · · ·   see under species
                                                           7e. *S.* × *intermedium*

S. aromaticum (*L.*) *Merr. & Perry* (*Caryophyllus aromaticus* L., *Eugenia aromatica* (L.) Baill., *non* O. Berg (U.O.P.Z.: 249 (1949)); *Eugenia caryophyllata* Thunb.; *Eugenia caryophyllus* (Spreng.) Bullock & S.G. Harrison; *Jambosa caryophyllus* (Spreng.) Nied. (T.T.C.L.: 377 (1949), Dale, Introd. Trees Uganda: 46 (1953)). Originally from the Moluccas, the clove tree was introduced into Zanzibar and Pemba in 1818 and rapidly became the major export accounting in the past for over 80% of the world's supply of cloves (the sun-dried flower buds) and clove oil. For more information see U.O.P.Z.: 249–252 (1949) and for an extensive history of the crop in East Africa, Stuhlmann, Beiträge zur Kulturgeschichte von Ostafrika (Deutsch-Ost-Afrika vol. 10: 278–295 (1909), as *Caryophyllus aromaticus*); it has also been cultivated at Amani (Tanzania. Lushoto District: Amani, 3 Feb. 1921, *Soleman A.H.* 6056 & 28 Oct. 1986, *Ruffo & Mmari* 2115 & Sigi Nursery, Lunguza, 23 Mar. 1973, *Ruffo* 1055; Zanzibar, Kanaoni, 31 Jan. 1929, *Greenway* 1252; Pemba, 13 Oct. 1929, *Burtt Davy* 22537).

Tree 5–21 m tall, much branched with a bushy conical crown. Leaves pink when young, rather shining, elliptic or oblong-elliptic, about 10 × 5 cm, very narrowly attenuate at the base, smelling of cloves when crushed. Flowers in terminal cymose clusters; unopened buds green, pinkish or brilliant crimson; opened flowers purplish, about 6 mm wide but petals soon falling. Stamens greyish yellow. Fruits dark purple, oblong-ellipsoid, 2–2.5 cm long, 1.3 cm wide, juicy, usually 1-seeded.

NOTE. The trees start flowering at about 7–8 years old and live for 60–70 years or more.

S. aqueum (*Burm.f.*) *Alston*, the 'wax jambo' (*Eugenia aquea* Burm.f., *E. javanica* Lam., U.O.P.Z.: 253 (1949)) has been grown in Tanzania (Lushoto District: Amani, 28 Oct. 1928, *Greenway* 938). Native of Sri Lanka, Andaman Is., Bangladesh, Burma and W Malaysia.

Tree up to 18(–23) m tall with short crooked ribbed bole and brown flaky bark. Leaves crimson when young, oblong to elliptic, 4.5–23 cm long, 1.5–11 cm wide, obtuse, retuse or shortly acuminate at the apex, cuneate, rounded or slightly subcordate at the base, coriaceous; petiole 1–5 mm long. Flowers white, ± 10 in terminal and axillary cymes, subsessile. Calyx-tube

1.5–3 cm long; lobes ovate, up to 6 mm wide. Petals pinkish white, oblong, up to 1.2 cm long, 8 mm wide, obtuse. Stamens pink, up to 1.5 cm long; anthers white. Fruit red, globose, 2 cm diameter, crowned with prominently necked persistent calyx-ring.

S. malaccense (*L.*) *Merr. & Perry*, the 'Malay apple' (*Eugenia malaccensis* L., U.O.P.Z.: 253 (1949); *Jambosa malaccensis* (L.) DC., T.T.C.L.: 378 (1949)) has been grown in Tanzania (Lushoto District: Amani, 28 Oct. 1928, *Greenway* 937, & Kiumba, 8 Mar. 1950, *Verdcourt* 102, & 20 Mar. 1973, *Ruffo* 668, & Sigi Hort. Station, 23 Mar. 1973, *Ruffo* 1070; Zanzibar (U.O.P.Z.) and doubtless many other places). Native of India, Malaya (U.O.P.Z.) and Polynesia (T.T.C.L.) now very widely cultivated in the tropics.

Conical tree 6–18 m tall with dense foliage and smooth flaky bark. Leaves very variable, oblong to elliptic, 16–34 cm long, 5–15 cm wide, acute or acuminate at the apex, cuneate at the base, subcoriaceous; petiole 0.5–1.5 cm long. Flowers crimson in ± sessile clusters on new or old wood. Calyx 1.5 cm long, 8 mm wide with lobes ± round, 5 mm long, 4 mm wide. Petals elliptic, 10 mm long, 8 mm wide, obtuse. Fruit white, yellow, reddish or pink-blotched, obovoid or oblong-pyriform, 2.5–6 cm long, scented, crowned with enlarged green sepals.

S. paniculatum *Gaertn.* (*Eugenia paniculata* (Gaertn.) Britten, *non* Lam. 1789 *nec* Jacq. 1788; *E. myrtifolia* Sims) has been cultivated around and in Nairobi as an ornamental (Nairobi, 23 May 1961, *Hemming* H121/61, & Karen, Hort. Gardner, 30 Dec. 1972, *Gillett* 20178, & Uhuru Ave., 22 Nov. 1974, *Gebreyesus* in *E.A.* 15527). Tree 6–15 m tall with grey bark. Leaves somewhat discolorous, elliptic to oblanceolate, 2.5–6 cm long, 1.3–2.7 cm wide, rounded to shortly acute at the apex, cuneate at the base; lateral nerves ± 30 on each side, prominent on both sides; petiole 3–5 mm long. Flowers few in terminal dichasia; pedicels 1–15 mm long. Calyx green with lobes tipped red but later all red; tube narrowly obconic, 5 mm long; lobes ± oblong, 3 mm long, 2.7 mm wide. Petals white, round, 4 mm long and wide. Stamens white, 6–8 mm long. Style 1.3 cm long. Fruit pink or carmine-red, ovoid, 1.3–1.5 cm long, 0.8–1 cm wide, fleshy, not eaten.

NOTE. Has been used as a street tree; cultivated up to 1800 m. Hyland (Austr. J. Bot. Suppl. 9 (1983)) distinguishes between *S. paniculatum* and *S. australe* (Link) B. Hyland which is the correct name for *Eugenia myrtifolia* Sims 1821 (*non* Salisb. 1796). He cites no specimens and despite going through much material at Kew, unfortunately none of it annotated by Hyland, and using his keys and descriptions I have not been able to determine to which species the Nairobi plants belong. There is not a single item in the description which does not overlap save the polyembryony of the seed in *S. paniculatum*. Unfortunately I have seen no seeds from the Nairobi plants.

1. **S. jambos** (*L.*) *Alston*, Handb. Fl. Ceylon 6 (Suppl.): 115 (1931); Amshoff in Fl. Gabon 11: 17 (1966); F. White in F.Z. 4: 204 (1978); Ashton, Rev. Handb. Fl. Ceylon 2:427 (1981); Thulin & Moggi in Fl. Somalia 1: 245 (1993). Type: Sri Lanka [Ceylon], *Hermann* 2: 20 (BM-Herb. Hermann!, lecto.*)

Much branched bushy glabrous evergreen tree, 7.5–15 m tall. Leaves lanceolate, 8–20 cm long, 1.3–5.5 cm wide, very acuminate at the apex, cuneate at the base, ± coriaceous; petiole 5–10 mm long. Flowers ± 5 cm in diameter, in terminal corymbs; pedicels 3–9 mm long. Calyx-lobes broadly rounded, 5–8 mm long, 8–12 mm wide. Petals white or yellowish, ± round, 1.5 cm diameter, concave. Stamens yellow, very conspicuous, 3–4 cm long. Style 4–5 cm long. Fruit white or pale yellow, rose-tinged, subglobose or pyriform, 2.5–5 cm long, 2–4 cm wide, 1-seeded, the flesh edible and rose-scented.

UGANDA. Cult. fide Dale
KENYA. Cult. fide Jex-Blake
TANZANIA. Mwanza District: probably Ukerewe I., 10 Mar. 1931, *Conrads* 6094!; Lushoto District: Mlalo Mission, Aug. 1955, *Semsei* 2212!; Kilosa District: Kilosa, Oct. 1952, *Semsei* 954!; Zanzibar: Jozani [Josani] Forest, Aug. 1972, *Robbins* 71!
DISTR. **U**; **K**; **T** 1, 3, 6; **Z**; native of tropical Asia now widely cultivated throughout the tropics and occasionally naturalised

* See Fawcett & Rendle, Fl. Jamaica 5: 352 (1926).

HAB. Cultivated but in Zanzibar undoubtedly naturalised in forest on loam over coral; cultivated 0–1350 m, only known to be naturalised at 10–20 m

SYN. *Eugenia jambos* L., Sp. Pl.: 470 (1753); U.O.P.Z.: 253 (1949); Popenoe, Man. Trop. Fruits: 305, t. 16 (1920); Jex-Blake, Gard. E. Afr. ed. 4: 112 (1957)
  Jambosa vulgaris DC., Prodr. 3: 286 (1828); Engl. & Nied. in P.O.A. C: 288 (1895), *nom. illegit.* Type: as for *S. jambos*
  J. jambos (L.) Millsp. in Publ. Field Mus. Nat. Hist., Bot. 2: 80 (1900); T.T.C.L.: 378 (1949); Dale, Introd. Trees Uganda: 46 (1953)

NOTE. This is certainly naturalised in Zanzibar at least; U.O.P.Z. states "more or less wild in the Protectorate". The fruit is mostly known as the 'rose apple' and although grown for its fruit, this is rather insipid and certainly not esteemed. Chang and Mizu describe a var. *linearilimbum* (Acta Bot. Yunnanica 4: 17 (1982)) with leaves 26 × 2.5 cm, solitary axillary flowers and narrower sepals. Since I have not seen this, the description above refers only to typical *S. jambos*.

2. **S. cumini** (*L.*) *Skeels* in U.S. Dept. Agric. Bur. Pl. Indust. Bull. 248: 25 (1912); Alston, Handb. Fl. Ceylon 6 (Suppl.): 116 (1931); T.T.C.L.: 379 (1949); Dale, Introd. Trees Uganda: 66 (1953); Amshoff in Fl. Gabon 11: 17 (1966); F. White in F.Z. 4: 203 (1978); Ashton in Rev. Handb. Fl. Ceylon 2: 443 (1981); J.F. Morton, Fruits Warm Climates: 375, pl. 52 (1987); K.T.S.L.: 127 (1994). Type: Sri Lanka [Ceylon], *Hermann* 1: 45 (righthand specimen) (BM-Herb. Hermann!, lecto.)

Glabrous evergreen tree or large shrub, 6–25 m tall with spreading crown; outer bark pale yellow brown, white or grey, densely thinly flaky (described as rough or smooth); inner bark rough, thick and fibrous. Leaves shiny, mostly very regularly oblong-elliptic but can be elliptic, oblanceolate-elliptic or ovate-lanceolate, 6–17(–20) cm long, 2.5–8.5 cm wide, acuminate to a rounded tip which is often down-curved, cuneate at the base; ± coriaceous; main lateral nerves numerous and close, ± 40; petiole 0.8–2 cm long. Cymes terminal and axillary, or mostly lateral borne in the axils of fallen leaves on older stems, about 10 cm long; flowers drying pale orange-brown. Calyx often drying orange-brown, obconic, 4–5 mm long, 3–4 mm wide, very obscurely 4(–5)-lobed, practically truncate, rim orange. Petals white, elliptic, 4 mm long, 3 mm wide. Stamens white, 4–6 mm long with very slender filaments. Style 5–6 mm long. Fruits dark purple-red, broadly ellipsoid, oblong or subglobose, often somewhat curved, 2.4–3.5(–5) cm long, 2 cm wide (2 × 1 cm dry), crowned by 2 mm diameter calyx-rim, 1-seeded (rarely 2–5-seeded or seedless in some varieties); pulp white, acid.

UGANDA. Mengo District: Kampala plantation, Apr. 1932, *Hopkins* in *Tothill* 1146!
KENYA. Nairobi, Eastleigh, 6 Jan. 1951, *Verdcourt* 410! & Eastleigh Section 7, near California Estate, 27 Nov. 1971, *Mwangangi* 1866!; Kilifi District: Jilore Forest Station, 28 Nov. 1969, *Perdue & Kibuwa* 10115!
TANZANIA. Bukoba District: Bukoba Forest Station, Munene Forest Reserve, 15 Oct. 1955, *Sangiwa* 82!; Pangani District: Boza, 9 Jan. 1956, *Tanner* 2541!; Zanzibar: near Chukwaní, 12 Feb. 1961, *Faulkner* 2752! & Pemba, Fufuni, 17 Dec. 1930, *Greenway* 2749!
DISTR. U 4; K 4, 7; T 1–4, 6; Z; P; India, Sri Lanka to S China, Malesia and Pacific; cultivated throughout tropics and often naturalised
HAB. Native plots, plantation edges, seashore and coastal bush and savanas on coral rock, more or less naturalised in some coastal areas; 0–1650 m

SYN. *Myrtus cumini* L., Sp. Pl.: 471 (1753)
  Eugenia jambolana Lam., Encyl. 3: 198 (1789); Wight, Ic. Pl. Ind. Or.: t. 535 (1843); Duthie in Fl. Br. India 2: 499 (1879); Trimen, Handb. Fl. Ceylon 2: 179 (1894); Popenhoe, Man. Trop. Fruits: 304 (1920); Jex-Blake, Gard. E. Afr. ed. 4: 301 (1957). Type: Jambolan, Rumph. Herb. Amboin. 1: 131, t. 42 (1741) (holo.)
  Syzygium jambolanum (Lam.) DC., Prodr. 3: 259 (1828)
  Eugenia cumini (L.) Druce in Rep. Bot. Exch. Club Brit. Is. [1913] 3: 418 (1914); Merr., Interpr. Rumph. Herb. Amboin.: 394 (1917); U.O.P.Z.: 252 (1949); Jex-Blake, Gard. E. Afr. ed. 4: 345 (1957)

Note. The bark is variably described as rough or smooth. Musk reports the tree reaches 100 ft in height but this seems excessive. The tree, often known as the Java plum or jambolan, is often confused with *S. guineense* in herbaria; the fruits are sold in the coastal markets. Linnaeus confused the specimens in Herb. Hermann as pointed out by Trimen in J.L.S. 24: 142 (1887). *M. cumini* L. is linked to Fl. Zeyl. 185 but the specimens in Hermann Herbarium 2: 82 numbered 139 and det. as *Jambolifera* are not Myrtaceae; the specimens in 1: 45 have 139 crossed out and 185 added in pencil.

3. **S. parvulum** *Mildbr.* in N.B.G.B. 14: 107 (1938); T.T.C.L.: 380 (1949); Amshoff in Acta Bot. Neerl. 8: 54 (1959). Type: Tanzania, NW Uluguru Mts, *Schlieben* 3922 (B†, holo., BM!, BR, Z, iso.)

Small tree or much-branched shrub 3–6 m tall; stems quadrangular when young with almost winged edges, the epidermis flaking on older branches. Leaves slightly discolorous, elliptic, 1.5–4.5 cm long, 0.9–2.4 cm wide, narrowed to a rounded apex above, cuneate at the base, subcoriaceous; midrib impressed above; venation closely reticulate above but becoming obscure; petiole 2–5 mm long. Inflorescences small terminal dichasial cymes 1.5–4 cm long; peduncle ± 1 cm long or absent when centre and lateral branches originate at terminal node; secondary branches ± 7 mm long; pedicels ± 1 mm long. Calyx-tube 1–1.5 mm long, lobes under 0.5 mm long. Petals greenish cream, 1.5 mm long, 1.8 mm wide, shortly broadly clawed. Stamens ± 15, ± 2 mm long. Style 1 mm long. Fruits red or plum-purple, subglobose, 0.9–1.2 cm long, 0.9–1.3 cm wide.

Tanzania. Morogoro District: Uluguru Mts, 4.8 km S of Bunduki, 15 Mar. 1953, *Drummond & Hemsley* 1615! & above Morningside, Bondwa Peak, May 1953, *Semsei* 1220! & Lukwangulu Plateau, 19 Sept. 1970, *Thulin & Mhoro* 1031!
Distr. **T** 6 (Uluguru); not known elsewhere
Hab. Montane forest including mist forest; 1900–2300 m

Note. Amshoff redescribed the species not realising that Mildbraed had done so using the same gathering.

4. **S. cordatum** *Krauss* in Flora 27: 425 (1844); Fl. Cap. 2: 521 (1862); Engl. & Nied. in P.O.A. C: 288 (1895); T.T.C.L.: 379 (1949); U.O.P.Z.: 458 (1949); Codd, Trees & Shrubs Kruger Nat. Park: 136, fig. 128/c (1951); I.T.U.: 273 (1952); Brenan in Mem. N.Y. Bot. Gard. 8: 439 (1954); Palgrave, Trees Centr. Afr.: 283 (1957); Amshoff in Acta Bot. Neerl. 9: 405 (1960); K.T.S.: 335 (1961); F.F.N.R.: 303 (1962); Boutique, F.C.B., Myrtaceae: 4, t. 1, fig. 1/a (1968); Amshoff in C.F.A. 4: 100 (1970); Palmer & Pitman, Trees S. Afr.: 1675, fig. & photo. (p. 1676) (1972); F. White in F.Z. 4: 193, t. 44 & 45/a (1978); Troupin in Fl. Rwanda 2: 494, fig. 156.1 (1983); Killick in Fl. Pl. Afr. 51: t. 2025 (1991); Lovett in E.A. Nat. Hist. Soc. Bull. 23 (4): 66–69 (1993); K.T.S.L.: 126, map (1994). Type: 'In sylvis dunarum per totam terram natalensem', South Africa, Durban [Port Natal], *Krauss* 136 (K!, iso.)

Evergreen tree or shrub (2–)3–20 m tall with dense spreading crown (occasionally in N Zululand a dwarf shrub only 35 cm tall); foliage often slightly glaucous; bark pale or dark brown, rough, fibrous and flaking or longitudinally fissured or reticulate; young stems quadrangular with acute angles or slightly winged or in variants scarcely quadrangular. Leaves oblong, oblong-elliptic, lanceolate-elliptic or almost round, 2.3–13.5 cm long, 2–7 cm wide, rounded to subacute or rarely very shortly acuminate or emarginate at the apex, cordate and amplexicaul or occasionally only subcordate, rounded or broadly cuneate at the base; petiole up to 2(–5) mm long. Flowers in dense dichasial cymes 5–7.5(–10) cm wide; axes ± square; pedicels 1–3 mm long. Calyx (tube and lobes) together with pseudopedicels 6–9 mm long. Calyx-lobes together with free part of calyx-tube 3.5–5 mm long. Petals yellowish or white, forming a calyptra up to 3.5 mm tall and 6 mm wide. Stamens cream or white, purple at base, (0.8–)1.1–1.5 cm long. Style (0.5–)1–1.7 cm long. Fruit mauve-purple or blackish purple, oblong, subglobose or urceolate, (0.8–)1.2–1.8 cm long, (0.5–)0.9–1.1 cm wide, bearing persistent calyx remains 3–4 mm long, 5 mm wide, edible.

subsp. **cordatum**

Young stems distinctly quadrangular or slightly winged. Leaves distinctly cordate and amplexicaul. Fig. 15.

UGANDA. Kigezi District: Rubanda, June 1939, *Purseglove* 739!; Elgon, Kaburon, Jan. 1936, *Eggeling* 2483!; Masaka District: NW side of Lake Nabugabo, 6 Oct. 1953, *Drummond & Hemsley* 4645! & 4645A!
KENYA. Trans-Nzoia District: near Moiben, 17 Jan. 1964, *Brunt* 1375!; Uasin Gishu District: Eldoret, 10 Apr. 1951, *Williams Sangai* 103!; Kisumu–Londiani District: Lumbwa to Londiani, 25 May 1953, *Verdcourt* 936!
TANZANIA. Musoma District: Klein's Camp, 6 Apr. 1961, *Greenway & Myles Turner* 10000!; Ulanga District: Upper Ruhudji R., Landschaft Upembe, 14 Sept. 1931, *Schlieben* 1190!; Rungwe District: near Tukuyu, K.A.R. H.Q., Masoko Crater, 21 Sept. 1936, *Burtt* 6300!
DISTR. U 1 (Debasien)–4; K 3–5; T 1, 4, 6–8; eastern and central Africa from Congo (Kinshasa) to Angola and South Africa
HAB. Swamp forest by lakes, sometimes in the water of *Papyrus* swamps etc., riverine woodland and forest, also *Brachystegia* and other open woodland (e.g. *Combretum* and *Acacia-Albizia*) and forest but still by streams etc.; 900–2400 m

SYN. *Eugenia cordata* (Krauss) Laws. in F.T.A. 2: 438 (1871); Hiern, Cat. Afr. Pl. Welw. 1: 360 (1898); Sim, For. Fl. Port. E. Afr.: 67 (1909)

subsp. **shimbaense** *Verdc.*, **subsp. nov.** a subsp. *cordato* surculis junioribus haud vel minus quadrangularibus, foliis basi subcordatis vel rotundatis haud distincte cordatis amplexicaulibus differt. Type: Kenya, Kwale District, Shimba Hills, Giriama Point, *Magogo & Glover* 28 (K!, holo., EA, iso.)

Medium-sized tree to 7 m tall; young parts of branches mostly squarish but sometimes almost imperceptibly so; bark fairly smooth, flaking off in small narrow strips. Leaves drying pale grey-green or brown, said to be glossy above in life, oblong-elliptic, obovate-elliptic or almost round, 2.5–10 cm long, 1.5–4.6 cm wide, rounded or retuse at apex, subcordate, rounded or broadly cuneate at the base; petioles 2–3(–5) mm long. Panicles small, up to 4–5 cm long, 5(–9) cm wide; peduncle and axes 1–2 cm long. Calyx as a whole including pseudopedicel 6–7 mm long; lobes purple-margined, broadly rounded, 0.8 mm long, 1.8 mm wide. Calyptra of petals white, 3.5 mm wide. Filaments white, up to 1.2 cm long. Style white, 1–1.7 cm long. Fruits slightly asymmetrically ellipsoid, narrowed at both ends, 1.8 cm long, 1 cm wide; fruiting pedicels 7 mm long.

KENYA. Kwale District: Shimba Hills, Kwale, *R.M. Graham* Q251 in F.D. 1775! & Shimba Hills Reserve, observation point, Giriama, 18 Nov. 1978, *Brenan et al.* 14559! & Longomwagandi area, 15 Feb. 1968, *Magogo & Glover* 104!
TANZANIA. Uzaramo District: Pugu road, 9.5 km [from Dar es Salaam], 12 Mar. 1939, *Vaughan* 2766!
DISTR. K 7; T 6; not known elsewhere
HAB. Forest edges, grassland and scattered trees, also open *Parinari* bushland; (<100–)300–450 m

SYN. ? *S. sclerophyllum* Brenan × *S. cordatum* Krauss; K.T.S.L.: 127 (1994)

NOTE. This population on the Shimba Hills is distinctive whatever its nature and it seems reasonable to name it. It has been suggested it is derived from hybrids between *S. cordatum* and *S. guineense* or *S. micklethwaitii* but none of these has been found in the area; it is true that Battiscombe recorded *S. guineense* from 'the coastal plains' but his specimen is unlocalised and resembles montane material. If the Shimba Hills taxon is of hybrid origin the parents would appear to have been hybridised out of existence locally. The variation in the leaf-base might be some evidence for hybrid origin but there is some variation of this character in undisputed populations of typical *S. cordatum*. The very close similarity between the Shimba population

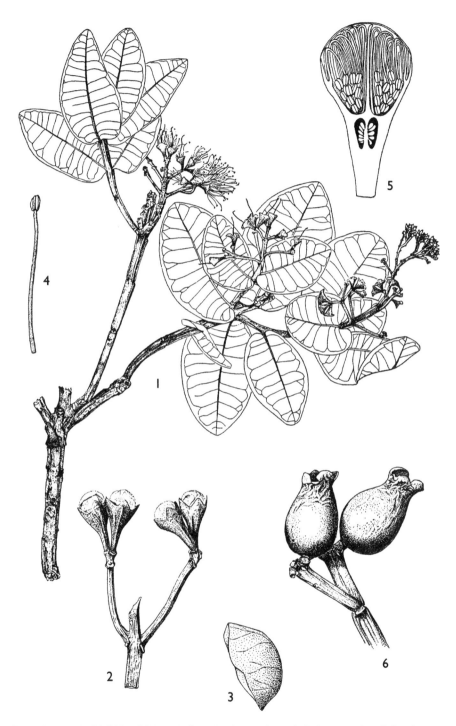

FIG. 15. *SYZYGIUM CORDATUM* — **1**, flowering branchlet, × ¹/₂; **2**, flower buds, × 2; **3**, calyptra, × 3; **4**, stamen, × 5; **5**, flower bud (longitudinal section), × 5; **6**, fruits × 2. 1–5, from *Devillé* 49; 6, from *Devillé* 485. Drawn by D. Leyniers. From F.C.B. Myrtaceae, t. I, reproduced with permission of the Nationale Plantentuin van Belgie.

and some South African material has persuaded me to look on it as derived from an isolated population of *S. cordatum.* The smoothly rounded leaf apices, often subcordate leaf-bases and short petioles all agree. No typical *S. cordatum* occurs so far as I know for some 400 km in any direction. Material from Tanzania has since been found at BM and is cited above.

5. **S. owariense** (*P. Beauv.*) *Benth.* in Hook., Niger Fl.: 359 (1849); Keay, F.W.T.A. ed. 2, 1: 240, t. 95c (1954); Amshoff in Acta Bot. Neerl. 9: 408 (1960); F.F.N.R.: 304 (1962); Amshoff in Fl. Gabon 11: 13 (1966); Boutique, F.C.B., Myrtaceae: 9, fig. 1/e (1968); Amshoff in C.F.A. 4: 109 (1970); F. White in F.Z. 4: 197 (1978). Type: Nigeria, Warri, *Beauvois* s.n.(BM!, lecto.*)

Tree (3–)9–30 m tall with straight bole and dense oblong bushy crown; bark grey, smooth or rough and flaky; young stems ± terete or subangular; stilt-roots or pneumatophores at least sometimes present. Leaves ovate to ovate-lanceolate, at least some broadest below the middle** and distinctly tapering from near the base, (4.5–)11–14 cm long, 2.4–6 cm wide, narrowly acuminate at the apex, rounded or truncate to broadly cuneate at the base, rather coriaceous; petiole (0.5–)1–1.4(–2.4) cm long, frequently twisted (fide Dawe). Inflorescences terminal and axillary, together (3–)7–18 cm long. Calyx and pseudopedicel together 4–6.5 mm long but calyx-tube is scarcely narrowed at base; calyx-lobes triangular, 0.2–0.4(–1) mm long, 1.5–1.6 mm wide. Petals purplish green, 1.5–2 mm long; filaments white, (0.5–)0.8–1.2 cm long. Style 5–7 mm long. Fruit purple-black, urceolate, (0.9–)1.2–1.8 cm long, (4–)9 mm wide.

UGANDA. Masaka District: S Buddu, ? Apr. 1905, *Dawe* 336!*** & *Dawe* 963!
TANZANIA. Mpanda District: Mpanda–Uvinza road, Uzondo Plateau, 30 May 2000, *Bidgood et al.* 4566!; Iringa District: Iheme Agricultural Station, 19 Oct. 1934, *Burtt* 6290! & near Sao Hill, 30 Oct. 1947, *Brenan & Greenway* 8242! & Sao Hill, Oct. 1935, *Mr & Mrs Hornby* 666!
DISTR. U 4; T 4, 7; Sierra Leone to Nigeria, Cameroon, Gabon, Congo (Kinshasa), ? Angola, Zambia, Malawi, Mozambique and Zimbabwe
HAB. Fringing forest in swampy places, particularly characteristic of true swamp forest, surrounding springs at sources of small perennial streams; 1100–2000 m

SYN. *Eugenia owariensis* P. Beauv., Fl. Owar. 2: 20, t. 70 (1810); F. Hoffm., Beitr. Kennt. Fl. Zentr.-Ost-Afr.: 36 (1889).
    Syzygium guineense (Willd.) DC. var. *palustre* Aubrév., Fl. For. Côte d'Ivoire ed. 1, 3: 70, t. 268/c (1936). Type: Ivory Coast, Dabou, *Chevalier* 1620 (P, holo.), *nom. invalid.* sine descr. lat.

NOTE. The distribution of this species is curiously erratic and little material is available from East Africa; Brenan had annotated this as *S. elegans* Vermoesen ined. (a name much used on labels but never published) before Keay showed the true identity of Beauvois's species. It has been much confused with *S. guineense* and the characters are not always obvious in herbarium material as Amshoff points out; several sheets are dubious and have been ignored in the distribution. The constancy of the presence of stilt roots needs investigation in the field.

6. **S. congolense** *Amshoff* in Acta Bot. Neerl. 9: 406 (1960) & in Fl. Gabon 11: 12, t. 2/1–5 (1966); Boutique, F.C.B., Myrtaceae: 10, fig. 1/f (1968). Type: Congo (Kinshasa), Zobia, *Claessens* 540 (BR, holo.)

Tree 6–30 m tall, sometimes buttressed, with a much-branched rounded habit and tendency to be flat-topped; young branches subterete to ± square, the angles often marked by a strong longitudinal rib decurrent from petiole-base; bark cream, silvery-grey or reddish brown, thin and smooth; slash dark brown, reddish or white and watery. Leaves drying characteristic yellowish green and discolorous, oblong, oblong-

---

* Keay also saw a leaf from a Beauvois specimen at G-DC; presumably there is a specimen at P
** Keay's figure hardly shows this
*** Dawe, Rep. Bot. Miss. Uganda Prot. (1906) gives only 262 which is not at Kew

elliptic or mostly obovate to obovate-oblong, 3.5–16 cm long, 1.5–6.5 cm wide, rounded but abruptly apiculate to shortly acuminate at the apex, cuneate at the base, ± revolute at the margin, often subcoriaceous; midrib strongly impressed above; petiole 9 mm long; young foliage reddish to ruby red. Inflorescences axillary and terminal, together 7–10 cm long and wide, the axes 4-angled. Flowers with buds 3–4 mm wide. Calyx and pseudopedicel 4–6 mm long; bracteoles obtriangular or obovate; calyx-lobes overlapping in bud, ± round, 1–2 mm long, 2–3 mm wide. Petals white, 2–3 mm diameter. Stamens white, 5–8 mm long. Style 5–8 mm long. Fruits violet or violet-purple, globular, 1–1.6 cm diameter.

UGANDA. Kigezi District: Kayonza, Apr. 1948, *Purseglove* 2653!; Mengo District: Entebbe, Kitubulu, July 1941, *Eggeling* 4410!; Masaka District: Sese Is., Kyewaga, 16 Nov. 1924, *Maitland* 547!
TANZANIA. Bukoba District: Rubare Forest Reserve, Oct. 1957, *Procter* 720! & same locality, July 1951, *Eggeling* 6232! & without exact locality, May 1924, *Bancroft* 188!
DISTR. U 2, 4; T 1; Cameroon, Gabon, Central African Republic, Congo (Kinshasa)
HAB. Rain-forest at lake edges; 1100–1500 m

SYN. [*S. guineense* sensu I.T.U. ed. 2: 273 (1952) pro parte, *non* (Willd.) DC.]

NOTE. There is no doubt in my mind that this is a distinct species. Brenan sorted it out in the herbarium as 'group I of Uganda' many years before Vermoesen distinguished it and Amshoff decided it was truly separable. The calyx character is small but correlated with a distinct facies in dried specimens; it can be instantly distinguished by anyone with a good eye. Two or three specimens from Kenya with very long leaves 14–15 × 3.5–3.8 cm have been tentatively referred to this species but almost certainly do not belong. The material is *Muchiri* 612 from Narok District, Endiabiri, source of R. Narosura, 12 Jan. 1981 and *Kerfoot* 2203, Kericho District, Belgut 1 Reserve, Cheptuiyet Forest, Aug. 1960 (said to = *Trapnell* 2325, a specimen I have not seen). Unfortunately the specimens are inadequate.

7. **S. guineense** (*Willd.*) *DC.*, Prodr. 3: 259 (1828); F.P.N.A.: 1: 667 (1948); T.T.C.L.: 380 (1949); Codd, Trees & Shrubs Kruger Nat. Park: 136, fig. 128/a–b (1951); Pardy in Rhodes. Agric. Journ.: 76, photo. (1952); I.T.U.: 273, t. 12, 60 (1952) pro parte; F.W.T.A. 1: 240 (1954); Brenan in Mem. N.Y. Bot. Gard. 8: 439 (1954); Palgrave, Trees Centr. Afr.: 287, t. & photo. (1957); Amshoff in Acta. Bot. Neerl. 9: 408 (1960); K.T.S.: 335, t. 20 (1961) pro parte; F.F.N.R.: 303 (1962); Amshoff in Fl. Gabon 11: 13 (1966); Friedr.-Holzh. in Prodr. Fl. SW.-Afr. 97: 1 (1966); Boutique, F.C.B., Myrtaceae: 13 (1968); Amshoff in C.F.A. 4: 104 (1970); Palmer & Pitman, Trees S. Afr.: 1681, fig. & photo. (p. 1682–1683) (1972); Hepper, W. Afr. Herb. Isert & Thonning: 80 (1976); F. White in F.Z. 4: 198 (1978); Thulin & Moggi in Fl. Somalia 1: 244, fig. 134 (1993); K.T.S.L., 127, fig. 72.4, map (1994); Friis in Fl. Eth. 2 (2): 77 (1996). Type: Benin [Dahomey], Sawi, *Isert* s.n. (B-W 9582, holo., C, iso., K!, photo.)

Trees, shrubs or pyrophytic subshrubs 0.2–30 m tall with smooth or rough bark; sometimes buttressed; young branches ± terete or somewhat 4-angled, mostly only strongly 4-angled in hybrids with *S. cordatum*. Leaves elliptic, oblong-elliptic or elliptic-lanceolate to obovate-elliptic, (3–)4–16 cm long, 1–7.7 cm wide, acute to acuminate to broadly rounded, subtruncate or even emarginate at the apex, cuneate at the base; petiole 0.6–2.5(–4.5) cm long. Inflorescence short to extensive, 5–19 cm long. Calyx-tube with pseudopedicels usually distinct. Filaments 3.5–9 mm long. Fruits variously red or purple, subglobose or ellipsoid, 0.8–3.5 cm long, 0.6–2.5 cm wide.

DISTR. Very widespread throughout tropical Africa from Senegal to Somalia and Arabia and south to S & SW Africa

NOTE. F. White states in F.Z. that there are 11 subspecies recognisable throughout Africa but does not seem to have listed them all anywhere. This species shows a greater range of variability than almost any other species in the tropical African flora. The infraspecific taxa are ill-defined and merge almost continuously yet are certainly convenient and constant over wide areas. The situation is complicated by hybridisation with *S. cordatum* and backcrossing again with both (see under × *intermedium* p. 81). I feel the rank variety would be more suitable but have retained subspecies, which has been used by recent authors, to avoid confusion. A great deal of fieldwork and use of modern techniques may unravel the meaning of the variation. The key is only a rough guide.

KEY TO INFRASPECIFIC VARIANTS AND HYBRID

1. Small pyrophytic subshrub typically 20–60 cm tall; petioles
     mostly short, (1–)2–5(–9) mm long (**T** 4, 7, 8) · · · · ·   d. subsp. *huillense*
   Shrubs to tall trees · 2
2. Fire-resistant shrub or tree of ± seasonally dry woodland
     or grassland with scattered trees; leaves 6.5–14 ×
     3.6–7.7 cm, often relatively wider than in other taxa,
     with apex broadly rounded, subtruncate, emarginate
     or acute to shortly acuminate; petioles usually well
     developed, 1–2.5(–4.5) cm long · · · · · · · · · · · · · ·   c. subsp. *macrocarpum*
   Mainly riverine or forest trees · 3
3. Leaves rounded at the apex or sometimes very bluntly
     acuminate, rounded to subcordate at the base; petiole
     often short, 2–6(–13) mm long · · · · · · · · · · · · · · ·   e. × *intermedium*
   Leaves acute or acuminate at the apex, cuneate at the
     base; petioles 0.7–1.3(–2.5) cm long · 4
4. Leaves distinctly acuminate; mostly evergreen forest · ·   b. subsp. *afromontanum*
   Leaves acute, subacute or indistinctly acuminate; mostly
     riverine ·   a. subsp. *guineense*

a. subsp. **guineense**; Boutique, F.C.B., Myrtaceae: 13, fig. 2/a (1968); Amshoff in C.F.A. 4: 105 (1970); F. White in F.Z. 4: 201, t.45/e (1978); Troupin, Fl. Rwanda 2: 496, fig. 156.3 (1983); K.T.S.L.: 127 (1994) pro parte; Friis in Fl. Eth. 2 (2): 77 (1996); Wood, Handb. Yemen Fl.: 173 (1997)

Slender tree 9–20 m tall with smooth to deeply fissured white, grey or brownish bark; young branches ± terete or not markedly 4-angled. Leaves elliptic, oblong-elliptic or elliptic-lanceolate, 3.7–13(–16) cm long, 1–5 cm wide, acute or shortly acuminate to subacuminate (but not so distinctly acuminate as in subsp. *afromontanum*) at the apex, cuneate at the base; petiole 0.7–1.2(–2.5) cm long. Calyx-tube with pseudopedicel usually quite slender. Petals 2–3 mm long. Stamens 5–6 mm long. Style 5–6 mm long. Fruits globose or ellipsoid, 8–12 mm long.

UGANDA. Mengo District: Buvuma I., 13 Mar. 1904, *Bagshawe* 614!; Mubende District: Kakumiro, 8 Mar. 1906, *Bagshawe* 953!
KENYA. W Suk District: Kacheliba, Suam R., *Thomas* 2110!; Trans-Nzoia District: Elgon, 5 Mar. 1935, *G. Taylor* 3847!
TANZANIA: Mwanza District: probably Ukerewe I., 8 Sept. 1931, *Father Conrads* 5978!; Arusha District: NE flank of Mt Meru, Dec. 1966, *Procter* 3426!; Mpanda District: Mokoloka, 19 Sept. 1958, *Newbould & Jefford* 2449!
DISTR. **U** 3, 4; **K** 2–4, 7; **T** 1, 2, 4, 6–8; widespread throughout tropical Africa from Senegal to Somalia to South Africa, also Saudi Arabia and Yemen
HAB. Essentially riverine forest and woodland of *Terminalia, Combretum, Acacia* etc.; 800–2050 m

SYN. *Calyptranthes guineensis* Willd., Sp. Pl. 2: 974 (1800)
     Syzygium guineense (Willd.) DC. var. *guineense* Keay, F.W.T.A. ed. 2, 1: 240, fig. 95/d, fig. 96 (1954); Amshoff in Fl. Gabon 11: 14 (1966)
     S. guineense (Willd.) DC. subsp. *guineense* Group A; F. White in F.Z. 4: 201, t. 45/e1–3 (1978)

NOTE. Since there is much confusion as to what typical *S. guineense* from W Africa is, it may be useful to give a short description – savanna tree ± 12 m tall (e.g. *Morton* GC 25261 from Ghana by road between Banda and Menji, Wenchi area), leaves elliptic, 7.3 × 3.3 cm (in type), very shortly acutely acuminate or acute at the apex, cuneate at the base; petiole 7 (type) –12 (GC 25261) mm long. Narrow part of calyx-tube (pseudopedicel) quite slender, 2

× 0.6 mm; lobes very broadly rounded, 0.5 mm long. Keay compared the *Isert* type from Berlin with material at Kew and matched it with *Anderson* 29 from Ghana, Volta R.

Much included here could be backcrosses between × *intermedium* and subsp. *guineense* or subsp. *afromontanum* but they usually have distinctly 4-angled young stems. Certainly many specimens accepted as typical *S. guineense* in South Africa are closer to plants I have called *S.* × *intermedium* than they are to typical W African subsp. *guineense*.

b. subsp. **afromontanum** *F. White*, F.F.N.R.: 455, 303 (1962); Boutique, F.C.B., Myrtaceae: 14, fig. 2/b (1968); Amshoff in C.F.A. 4: 107 (1970); Chapman & F. White, Evergreen Forests Malawi: 44, photo. 21 (1970); F. White in F.Z. 4: 199, t. 45/f (1978); K.T.S.L.: 127 (1994); Friis in Fl. Eth. 2 (2): 78 (1996). Type: Zambia, Ndola, *White* 3058 (K!, holo., FHO, iso.)

Tree (3–)9–30 m tall, sometimes buttressed for 1.8 m, with rough or more usually smooth light grey to dark brown or white bark; slash tan, becoming white inside; young shoots usually ± terete or sometimes ± 4-angled. Leaves narrowly to broadly elliptic, 5.5–12(–14) cm long, 2.2–5(–6) cm wide, typically sharply narrow acuminate (rather than just acute or narrowed) at apex, cuneate or sometimes rounded at the base; petiole 0.6–1.3 cm long. Calyx-tube with pseudopedicels evident or often not evident in intermediates with *S. masukuense*. Fruit purple to almost black when ripe, obovoid or subglobose, 2–2.5 cm long, 0.6–1.3 cm wide.

UGANDA. Acholi District: Imatong Mts, Apr. 1938, *Eggeling* 3568!; Kigezi District: Impenetrable Forest, Oct. 1940, *Eggeling* 4174!; Mbale District: Elgon, Bulago, 22 Apr. 1927, *Snowden* 1080!
KENYA. Elgon, Feb. 1938, *Napier* 2528!; Machakos District: summit of Ol Doinyo Sapuk, 2 July 1967, *Gillett* 18306A!; Kericho District: Kericho, 13 May 1955, *Nicholson* 70!
TANZANIA. Lushoto District: E Usambaras, Ngambo, 11 Dec. 1940, *Greenway* 6077!; Kondoa District: near Bereku, Saranka Forest, 19 June 1973, *Ruffo* 762!; Iringa District, near Sao Hill, 1 Aug. 1933, *Greenway* 3438!
DISTR. **U** 1–3; **K** 1/2, 3–5, 7 (see note); **T** 2–8; from Sudan to Congo (Kinshasa) south to Angola, Mozambique and Zimbabwe
HAB. Evergreen forest slopes, escarpment crests, streamsides but by no means always riverine, forest edges also in *Acacia, Combretum, Brachystegia* woodland e.g. on Bereku Escarpment; (450–)900–2250 m (see note)
SYN. [*S. guineense* sensu K.T.S., t. 20 (on p. 335) (1961), *non* (Willd.) DC. sensu stricto]
NOTE. White's type of this subspecies is 'not typical' of the afromontane taxon as it occurs in the upland evergreen forests; in fact it seems to me nearer the type of *S. guineense* itself. It would be foolish to tamper with the names in common use until the whole complex has been examined throughout Africa by modern methods so I have left matters as they are. Certainly the afromontane populations are different from W African coastal populations.

Purseglove 866 (Uganda, Ankole, Ijara at 1650 m) growing in short grassland (derived from forest?) is stated to be only 1.2 m tall.

Subsp. *afromontanum* merges with subsp. *guineense* but there are specimens from lowland areas which have narrowly acuminate leaves e.g. *Vaughan* 2734 (Tanzania, Uzaramo District, km 16 Utete road (from Dar es Salaam), 21 Jan. 1939 at under 50 m which has slender petioles 1.7 cm long and slender pseudopedicels). *Battiscombe* 120 seems indistinguishable from subsp. *afromontanum* and is stated to be 'common tree on grassland in Coast District' but no locality is given. Battiscombe, T.S.K. ed. 1: 20 (1926) states 'also a tall tree of the coast district'.

Subsp. *afromontanum* also merges or crosses with *S. masukuense* and in the Mbizi Forest, 2180 m, Ufipa District, Tanzania the leaves seem to be rather more coriaceous and the pseudopedicels scarcely developed (e.g. *Whellan* 1359); despite the ± narrowly cuneate leaf-bases one suspects some influence from *S. masukuense*.

c. subsp. **macrocarpum** (*Engl.*) *F. White* in F.F.N.R.: 455, 304 (1962); Boutique, F.C.B., Myrtaceae: 16, fig. 2/e (1968); Amshoff in C.F.A. 4: 106 (1970); Friis in Fl. Eth. 2 (2): 78 (1996). Type: no specimens cited but many general localities given; lecto/neotype: Cameroon, Lom R., Kongola, *Mildbraed* 9076 (K!, lecto., chosen here)*

* Engler included Cameroon in his list of localities and would have seen this number. Mildbraed (Z.A.E. II 1910–11 (2): 10 (1922)) cites it as *S. campicola* ("campicolum") but this is a *nomen nudum*.

Fire-resistant tree or shrub 3–12 m tall; bark silvery-grey or grey-brown, variously described as fibrous, irregularly long-flaking, reticulate, rough or smooth but predominantly rough; slash dark chestnut. Leaves mostly larger and relatively broader than in other varieties, elliptic to obovate-elliptic, 6.5–14(–15) cm long, 3.6–7.7 cm wide, broadly rounded, subtruncate, emarginate or acute to shortly acuminate at the apex; petiole usually well developed, 1–2.5(–4.5) cm long. Inflorescences usually extensive, the terminal and axillary combined, up to 12–19 cm long. Fruits purple, mostly ± globular, 1.2–3 cm diameter.

Uganda. West Nile District: Otze Forest Reserve, Dec. 1947, *Dale* U493!; Bunyoro District: Chiope, *Dawe* 837!; Busoga District: Bulimwogi, *Snowden* 187!
Kenya. Trans-Nzoia District: Kitale, 5 Mar. 1953, *Bogdan*, 3655! & Kirks Bridge, 18 Apr. 1943, *Bally* 2508! & Kitale, 27 Feb. 1953, *Brockington* 40!
Tanzania. Mwanza District: Geita, Sept. 1949, *Watkins* 303!; Mpanda District: 16 km E of Mpanda, 22 July 1961, *Boaler* 292!; Singida District: Singida, rift wall above Mgori, Oct. 1935, *Burtt* 5236!
Distr. U 1–3; K 3; T 1, 4–8; West Africa to Cameroon, Gabon, Sudan, Central Africa and Congo (Kinshasa)
Hab. Woodland of *Julbernardia–Isoberlinia*, *Brachystegia–Uapaca*, *Combretum–Terminalia* etc. and seasonally wet grassland with scattered trees, *Protea*, *Combretum* etc., sometimes by water but even when not, underground water is present; 900–1860 m

Syn. [*Eugenia owariensis* sensu Lawson in F.T.A. 2: 438 (1871) pro parte, *non* P. Beauv.]
    Syzygium guineense (Willd.) DC. var. *macrocarpum* Engl. in V.E. 3 (2): 738 (1921); Aubrév., Fl. For. Soud-Guin.: 89, t. 14 (1950); F.W.T.A. ed. 2: 241 (1954); K.T.S. 335 (1961); Amshoff in Fl. Gabon 11: 14 (1966), as '*macrocarpon*'
    [*S. mumbwaense* sensu Dale, T.S.K. ed. 2: 29 (1936), *non* Greenway]
    [*S. owariense* sensu auctt. mult. e.g. I.T.U. ed. 2: 275 (1952), *non* (P.Beauv.) Benth.]
    S. guineense (Willd.) DC. subsp. *guineense* Group B; F. White in F.Z. 4: 202, t. 45/e4–8 (1978)

Note. Although recognisable over a wide area, some specimens do have leaves as narrow as 5.5 × 2.2 cm; it certainly merges with other subspecies and White ceased to recognise it in his F.Z. account. *Jackson* 354 (from Elgon, said to be 24–30 m tall and from 2040–2100 m) is puzzling; it looks like subsp. *macrocarpum*.

    d. subsp. **huillense** (*Hiern*) *F. White*, F.F.N.R.: 504, 304 (1963); Boutique, F.C.B. Myrtaceae: 17, fig. 2/f (1968) pro parte; F. White in F.Z. 4: 202, t. 45/h (1978). Type: Angola, Huila, between Mampula and Nene, *Welwitsch* 4401 (BM!, lecto., COI, K!, LISU, isolecto., chosen by Amshoff in Acta Bot. Neerl. 9: 406, 1960)

Subshrub (geopyrophyte) with woody annual shoots, typically (20–)45–60 cm tall (but merging with and otherwise indistinguishable from plants 0.9–4.5 m tall) from a woody underground rootstock. Leaves often drying a pale greyish brown, elliptic to obovate, 4–15.5 cm long, 2–7.5 cm wide, subacute, rounded acute to round or retuse at the apex, cuneate to round or almost cordate at the base; petiole usually short, (1–)2–5(–9) mm long. Calyces usually dry an orange-brown. Fruit purple or plum-coloured, 1.5–2(–3) cm long and wide.

Tanzania. Mpanda District: 58 km S of Uvinsa, 30 Aug. 1950, *Bullock* 3265!; Njombe District: near Luwangu Mission, 10 Nov. 1960, *Willan* 541! & Njombe, 10 Dec. 1931, *Lynes* C6!
Distr. T 4, 7, 8 (intermediates); Congo (Kinshasa), Angola, Zambia, Zimbabwe
Hab. Grassland, grassland with scattered trees, *Parinari*, *Protea*, *Erythrina* etc.; (1050–) 1450–1800 m

Syn. *Eugenia guineense* (Willd.) Laws. var. *huillense* Hiern, Cat. Afr. Pl. Welw. 1: 359 (1898) pro parte
    Syzygium huillense (Hiern) Engl. in E.J. 54: 339 (1917); Milne-Redh. in K.B. 2: 24 (1947); Amshoff in Acta Bot. Neerl. 9: 406 (1960) & in C.F.A. 4: 101 (1970)
    S. mumbwaense Greenway in K.B. 1928: 196 (1928) pro parte excl. *Bourne* 91. Type: Zambia, Mumbwa, Chibuluma vlei, *Macaulay* 995 (K!, holo.)

Note. In East Africa there is little to support keeping this subspecies separate; small subshrubs are indistinguishable from 7.5 m trees save for size e.g. *Davies* 23 from Mbosi and *Napper* 877 from the Mbisi Forest; but over much of Congo (Kinshasa), Angola and Zambia subsp. *huillense* is apparently constant and distinct. As in so many Central African species comprising forest and savanna paired taxa where, despite of, or perhaps even because of, the constant advance and retreat of the evergreen forest and savanna interface due to climatic changes over several million years, there is still gene flow between the tall tree types of 30 m and the savanna pyrophytic subshrubs of 30 cm; it is very often impossible to distinguish herbarium specimens without habitat data. It is as inconvenient to try separating the two as it is silly to pretend that the two are not different. This is particularly the case when the forest form is a good timber tree and the savanna form is little more than a herb. Common sense demands that some name at some rank is available. In other cases the separation of the taxa is complete.

e. **S. × intermedium** *Engl. & Brehmer* in E.J. 54: 339 (1917); V.E. 3 (2): 736 (1921); T.T.C.L.: 380 (1949); Boutique, F.C.B., Myrtaceae: 12, fig. 1/h (1968). Type: South Africa, Natal, Port Shepstone District (Alexandra County), Dumisa area, Friedenau, Umgayeflat, *Rudatis* 1174 (BM!, lecto.*, chosen here)

Small to medium tree or shrub (1.8–)4–15 m tall with large crown and mostly reticulate rough bark; young stems often 4-angled or slightly winged. Leaves often glaucous, lanceolate-elliptic to oblanceolate-elliptic, 7–15 cm long, 3.5–6.5 cm wide, rounded to bluntly acute, very shortly obtusely subacuminate or even retuse at the apex, broadly rounded to subcordate at the base; petiole usually short, 2–6(–13) mm long. Calyx including pseudopedicels 4.5–6 mm long; pseudopedicels mostly slender. Filaments 0.8–1.2 cm long. Style 0.5–1.3 cm long. Fruit red to purple-black, urceolate, up to 2 cm long, 1–1.6 cm wide, crowned with persistent calyx.

UGANDA. Mbale District: Bugishu, between Binyinyi and Kaburon, Jan. 1948, *Eggeling* 5738! & Elgon, Sebei, 18 Feb. 1924, *Snowden* 829!
KENYA. Northern Frontier Province: Mathews Range, 4 Sept. 1977, *Ichikawa* 786!; Nairobi, near Arboretum, Nov. 1940, *Mrs Bally* in *Bally* 1245!; Masai District: Siyabei Gorge where it cuts the road 21 km from Narok, 11 Dec. 1963, *Verdcourt* 3818!
TANZANIA. Morogoro District: Turiani, Manyangu Forest, Nov. 1953, *Semsei* 1412!; Iringa District: Mtwivila, 12 Nov. 1980, *Ruffo* 1541!; Lindi District: Lake Lutamba, 12 Sept. 1934, *Schlieben* 5326!
DISTR. U 3; K 1, 4–6; T 2–8; south to South Africa, Congo (Kinshasa), Angola, Zambia, Malawi, Mozambique and Botswana
HAB. Riverine, evergreen forest and woodland; 1050–2100 m

SYN. *S. deiningeri* Engl. in E.J. 54: 340 (1917); V.E. 3 (2): 737 (1921); T.T.C.L.: 379 (1949). Type: Tanzania, W Usambaras, Lushoto [Wilhelmstal], Jägertal, *Deininger* in *Holtz* 2753 & 2880 (B†, syn.)
  S. cordatum × *S. guineense*, F. White in F.Z. 4: 194, t. 45/b (1978)

NOTE. Earlier mentions of the name by Engler in Sitz. Preuss. Akad. Wiss. Berlin 52: 888 (1906) and R.E. Fries, Wiss. Ergebn. Schwed. Rhod.-Kongo-Exped. 1: 177 (1914) are nomina nuda or illegitimate.
  Certainly some of the syntypes of *S. × intermedium* are simple hybrids and the lectotype probably is, but the central problem is the frequent total lack of the parents in areas where apparent hybrids occur. This may mean there are separate taxa of *Syzygium* very similar to the hybrids in morphology or that the hybridisation has been so extensive that the parents have been hybridised out of existence. The populations vary considerably possibly explained by hybrid backcrossing with the parents. White mentions that in the Victoria Falls forests hybrids greatly outnumber the parents. Hybrids are fertile and the fruit is of course eaten by birds and mammals so distribution takes place rapidly. These problems are not soluble in the herbarium and there is immense scope for research here using modern techniques. Certain populations are mentioned in more detail below:
  • W Usambaras, Lushoto area: near old German Eucalyptus plantation, 12 May 1947, *Boniface* in *Hughes* 39; Lushoto, 10 June 1953, *Drummond & Hemsley* 2876; Kwengoma public lands, 28 Nov. 1968, *Ngoundai* 149; mostly wet sites in valley bottoms or swampy meadows by small streams at 1200–1400 m. Tree 12–17 m tall with grey or dark brown scaly bark; trunk sometimes divided into three. Leaves slightly glaucous, mostly large, elliptic-oblong to

* Since this is the only specimen of a cited number I have seen it seems the best choice for lectotype. Other syntypes were Tanzania; "Sansibar Kustenland bei Mucha und Kassanga" at 1000 m, *Holtz* 2231; gallery forest, Kilimanjaro area, Mbaschi at 1100 m, *Endlich* 99 and Zimbabwe, Victoria Falls at 930 m, *Engler* 2961, 2965a; all destroyed at B.

oblong, 8–17 cm long, 4–6 cm wide, very shortly obtusely acuminate to obtuse at the apex, rounded to broadly cuneate or ± truncate at the base; petiole 6–8 mm. Style 1.3 cm long. This description agrees quite well with that of *S. deiningeri* described from swampy ground at 1450 m near Lushoto with leaves 7–10 × 3.5–5 cm, save in one character – *S. deiningeri* is said to have slightly hairy leaves in the original description but this is *not* repeated by Engler in V.E. 3 (2): 737 (1921). *Drummond & Hemsley* make a special mention of absence of indumentum in the plants they examined clearly with this in mind although they do not mention *deiningeri*. I think that the hairs in the original material were probably fungal hyphae. *S. cordatum* has not been found in the W Usambaras.

• Tabora District, Ugalla and Wala R. area: Ugalla R., Isimbila, 25 Oct. 1960, *Richards* 13390!; Tabora, *C.H.N. Jackson* 14; Wala R., Sept. 1937, *Lindeman* 437; riverine woodland, ± 1080 m. Shrub or small tree 4–5 m tall. Leaves drying reddish brown, oblong-elliptic, 4–9 cm long, 2.5–3.8 cm wide, rounded to very bluntly shortly acuminate at the apex, distinctly cuneate to broadly rounded at the base; petiole 5–10 mm long. Style 5–6 mm long. This is a distinctive plant known from a number of specimens collected over many years. It certainly is not a simple hybrid of *cordatum* and *guineense* and might be a distinct subspecies of the latter. The area does have a number of endemic taxa.

• *Bally* 9872 (Kenya, Masai District, Amboseli, 14 Sept. 1954) is described as having the base of the trunk submerged in a swamp; no other specimens have been seen from there.

• It is interesting to note that from Gan Libah in Somali Republic (N) (e.g. *Glover & Gilliland* 1161) near a waterhole at 1260 m material very close to *S.* × *intermedium* is found. The possibility that the *S. cordatum* influence might have stretched so far is remarkable. Material of the species from Saudi Arabia and the Yemen is more evidently subsp. *guineense*. *S. guineense* subsp. *guineense* would have been present in the Sahara when it was much wetter and migration across might have been easy. *S. cordatum* certainly extended up the eastern side of Africa and could have reached further north than now.

f. Unplaced variants

Milne-Redhead & Taylor 8366 (Tanzania, Songea District, about 32 km E of Songea by R. Mkurira, 19 Jan. 1956, in riverine forest at 930 m) is a very distinctive variant. Evergreen shrub to 9 m. Leaves elliptic up to 12 × 5 cm, long-cuspidate at the apex, the tip narrow up to 2 cm long, 2–3 mm wide, narrowly attenuate into a petiole up to 2 cm long. Fruits turning through purple to black, subglobose, 1 cm diameter when dry.

Styemstedt TRU 527 (Tanzania, Tunduru District, 56 km NW of Tunduru, Msamala, Jan. 1967, in open miombo woodland at 750 m) is somewhat similar. The long petioles suggest subsp. *macrocarpum* but even if not rounded the leaf tips in this subspecies are not so long-acuminate.

8. **S. masukuense** (*Baker*) *R.E. Fr.*, Wiss. Ergebn. Schwed. Rhod.-Kongo-Exped. 1: 177 (1914); Brenan in Mem. N.Y. Bot. Gard. 8: 439 (1954); F. White in F.Z. 4: 196, t. 45/c–d (1978). Type: Malawi, Misuku Hills [Masuku Plateau], *Whyte* (K!, holo.)

Evergreen tree to 20 m tall or shrub 0.9–2 m tall with often wide spreading crown and dark brown rough scaly bark or trunk sometimes smooth; young stems often quadrangular or slightly winged but sometimes only obscurely angular. Leaves elliptic to elliptic-lanceolate or oblong-elliptic to almost round, 1.5–11 cm long, 0.9–5 cm wide, distinctly acuminate at the apex or in subsp. *pachyphyllum* cuspidate or shortly acuminate, cuneate, broadly rounded or subcordate at the base; petiole 1–8(–10) mm long. Inflorescence up to 6 cm long including peduncle, up to 8 cm wide. Calyx together with pseudopedicel 3–6 cm long, calyx-tube 3–4 mm long; lobes 0.5–1 mm long. Filaments 3–8 mm long. Fruit purple-black, subglobose, 1.2–1.5 cm long and wide, crowned by persistent calyx.

subsp. **masukuense**; F. White in F.Z. 4: 197, t. 45/c (1978)

Tree 6–15 m tall, rarely shrubby; slash brown, thick and fibrous. Leaves lanceolate-elliptic or elliptic, 4.5–11 cm long, 2.5–5 cm wide, distinctly acuminate at the apex, broadly rounded to subcordate at the base; petiole 2–8(–10) mm long. Calyx together with pseudopedicel 4.5–6 mm long; corolla cream. Filaments 2.5–8 mm long.

TANZANIA. Iringa District: Mufindi Tea Estate golf course, 27 Apr. 1984, *Lovett & Congdon* 282! & Mufindi Club Course, near 5th green, 3 Sept. 1971, *Perdue & Kibuwa* 11380! & Sao Hill, Iporogoro–Mkawa track, 12 Dec. 1961, *Richards* 15558!

DISTR. **T** 7; Malawi, Zimbabwe

HAB. Relict forest of *Trichocladus, Garcinia, Mystroxylon, Cassipourea* etc.; 1500–1950 m

SYN. *Eugenia masukuensis* Baker in K.B. 1897: 267 (1897)

NOTE. *Richards* 15558 had been determined as the W African submontane species *S. staudtii* (Engl.) Mildbr. to which it shows much resemblance but I have seen no fruits of *S. staudtii*. *S. masukuense* is much the older name. The problem of the relationship of *S. masukuense* and its small-leaved sometimes dwarf subsp. *pachyphyllum* F. White to *S. micklethwaitii* Verdc. and its small-leaved subspecies is one I have not attempted to clear up in this flora. The obvious solution is to combine them and have all as subsp. of *S. masukuense*. Really the only difference is the spectrum of leaf-shape – in *S. masukuense* the leaves are narrower with distinctly acuminate apices whereas in *S. micklethwaitii* they are broader, more rounded, with rounded apices. The type of subsp. *pachyphyllum* had originally been determined as *S. sclerophyllum* Brenan (now in synonymy of *S. micklethwaitii* subsp. *micklethwaitii*); it is curiously variable in habit from a small shrub 0.9 m tall to a tree 12–20 m tall. Without much more field work I have hesitated to sink the two taxa, particularly as other montane taxa are probably involved. F. White (F.Z. 4: 197 (1978)) commented on the problem and gave reasons for considering subsp. *pachyphyllum* distinct from *S. sclerophyllum*. Drummond & Hemsley 1985 (Morogoro District, S Ngurus, Ruhamba Peak, 2 Apr. 1953, in *Allanblackia, Podocarpus, Ocotea* forest) could equally be placed in either *S. micklethwaitii* or *S. masukuense*. D.G.B. Leakey in F.D. 3330 (Kenya, Laikipia), a tree to 15 m with young shoots markedly quadrangular, leaves oblong, 11.5 × 5 cm, distinctly acuminate at apex, rounded at base, drying brown, petiole 7 mm and young fruits elongate had been named *S. masukuense* by Amshoff – it is probably a hybrid of *S. guineense* with *S. cordatum*.

9. **S. micklethwaitii** *Verdc.* in K.B. 52: 682 (1997). Type: Tanzania, W Usambaras, Magamba, *Wigg* 88 (K!, holo., EA!, FHO, iso.)

Small to fairly large tree or occasionally shrubby, (1.5–)3–25 m tall, typically with very dense bushy rounded profusely branched crown, glabrous; branches 4-angled, the angles acute or almost winged, 2–7 mm wide; bark grey or grey-brown, slightly reticulate or rough. Leaves pinkish green when young, elliptic, broadly elliptic or almost round, 1–6(–8) cm long, 1–4(–6) cm wide (see note), rounded or less often subacute or shortly acuminate at the apex, broadly cuneate to rounded or slightly cordate at base, sometimes shortly decurrent into the petiole, coriaceous; lateral nerves 12–15, costa deeply impressed above, venation reticulate; petiole 1–5 mm long, often as wide as long, rugose, channelled above. Inflorescences terminal, sessile with no true peduncle or less often pedunculate, obpyramidal-corymbose or ovoid, (1–)2.5–4 cm long, (1.5–)3–5(–7) cm wide, usually trichotomous at base; primary axes 4-angled, typically 1–2 cm long, 1.2–3 mm wide. Flowers pinkish, in glomerules, subsessile, the calyx-tube basally scarcely developed into a pseudopedicel; buds 6 mm long, 4 mm wide. Calyx-tube 6 mm long, 4–5 mm wide at apex, slightly angled, produced above into a 3 mm cup; lobes 4, purplish, broadly rounded, 1.7 mm long, 2–2.5 mm wide at base. Petals 4, ± free (?), pale green or purplish green, concave, ± round, 4.5 mm long, 5 mm wide, soon falling. Stamens numerous; filaments white, 6–8 mm long; anthers 1 mm long, with small red gland below apex of connective. Style lilac-pink, 6 mm long. Fruit purple, subglobose, 0.8–1.6 cm long, 0.7–1.4 cm wide, crowned by calyx-lobes.

subsp. **micklethwaitii**

Leaves elliptic with ± slender petioles or if round or subcordate and ± sessile then petiole remnant narrow.

KENYA. Teita District: Taita Hills, Mbololo Hill, 31 Dec. 1971, *Faden et al.* 71/1002 & 8 km NNE of Ngerenyi, Ngangao, 15 Sept. 1953, *Drummond & Hemsley* 4360! & summit of Sagalla Hill, 4 Feb. 1953, *Bally* 8711!

TANZANIA. Lushoto District: W Usambaras, between Magamba and Gologolo, 6 Nov. 1947, *Brenan & Greenway* 8295! & 8296! & Matondwe, near summit of mountain above Kwai, 29 May 1953, *Drummond & Hemsley* 2811!; Morogoro District: S Ngurus, Ruhamba peak, 2 Apr. 1953, *Drummond & Hemsley* 1985!

DISTR. **K** 7; **T** 3, 6; not known elsewhere

HAB. Rather dry evergreen forest, mist forest, stunted forest on steep rocky slopes with *Podocarpus, Ficalhoa, Macaranga* etc., also transitional areas between upland forest and heath; 1350(Kenya)–2250 m

SYN. *S. sclerophyllum* Brenan in K.B. 4: 79 (1949) & T.T.C.L.: 380 (1949); Beentje in Utafiti 1: 50 (1988) & K.T.S.L.: 127 (1994), *non* Thwaites

NOTE. *Mabberley & Salehe* 1495 (Kilosa District, Ukaguru Mts, Mamiwa Forest Reserve, summit of Mamiwa, 16 Aug. 1972, in mist forest at 2310 m – a dominant tree with *Polyscias*) has to be regarded as a form of subsp. *micklethwaitii* but the inflorescence is pedunculate and has longer axes. *Mgaza* 611 (W Usambaras, Shagayu Forest, 30 Sept. 1964) with leaves up to 13.5 cm long, 8.5 cm wide is probably from a sapling or sucker shoot. There is great variation in the prominence of the venation in herbarium specimens from obscure to clear-cut and very prominent, correlated with a dull to very shiny surface e.g. in *Faden et al.* 386 from Kenya, Mbololo Hill, 13 May 1985.

subsp. **subcordatum** *Verdc.*, **subsp. nov.** a subsp. *micklethwaitii* foliis subrotundatis vel obovato-rotundatis, apice rotundatis vel retusis raro subacutis, basi subcordatis, petiolis brevissimis differt. Typus: Tanzania, Uluguru N Forest Reserve, Bondwa Ridge, *Mabberley* 1138 (K!, holo.)

Leaves almost round, 1–5.5 cm long and wide, rounded, retuse or rarely (in Ukaguru Mts) ± subacute at the apex, subcordate at the base; petioles very short, no longer than broad. Inflorescences congested.

var. **subcordatum**

Stunted tree 3–10 m tall with even young shoots 4–7 mm thick when dry. Leaves 3–5.5 cm long, 2–5.5 cm wide, broadly rounded or often truncate or retuse at the apex; petioles 3–5 mm wide. Flowers very pale pink.

TANZANIA. Morogoro District: Uluguru Mts, Lupanga Peak, Apr. 1935, *Bruce* 999! & W margin of Lukwangule Plateau, above Chenzema, 6 Dec. 1969, *Harris et al.* 3697! & edge of Lukwangule Plateau, 17 Mar. 1953, *Drummond & Hemsley* 1664!

DISTR. **T** 6; not known elsewhere

HAB. Elfin forest with *Allanblackia, Philippia, Lobelia* etc., short grassland on the timber line and poorly developed upland rain forest with *Rapanea*, dwarf *Ocotea, Trichocladus* and *Ternstroemia*; 2100–2450 m

NOTE. Mildbraed has annotated *Schlieben* 3551 from W Lukwangule, 22 Feb. 1933 as *S. cordatum* var. *alticola* Mildbr. n. var. but I can find no publication of this.

var. **dryas** *Verdc.* **var. nov.**, a var. *subcordato* surculis junioribus gracilioribus, in siccitate 2–3 mm crassis; foliis minoribus 1–2.5(–3 cm) longis latisque; petiolis 1–1.5 mm latis differt. Typus: Tanzania, Uluguru Mts, W part of Lukwangule Plateau, *Harris et al.* 3725 (K!, holo., DSM, EA, iso.)

Shrub or small tree 3–7 m tall; young shoots slender, only 2–3 mm thick when dry. Leaves 1–2.5(–3) cm long and wide, broadly rounded at apex (see note); petioles 1–1.5 mm wide.

TANZANIA. Lushoto District: W Usambaras, Shagayu Peak, 24 May 1953, *Drummond & Hemsley* 2732A!; Morogoro District: Lukwangule Plateau, 19 Sept. 1970, *Thulin & Mhoro* 997!

DISTR. **T** 3, 6; not known elsewhere

HAB. Subalpine forest/elfin woodland, stunted forest with *Balthasaria* (*Melchiora*), *Schefflera* and *Podocarpus*; 2200–2500 m

NOTE. *Mabberley et al.* 1286 (Kilosa District, Ukaguru Mts, Mamiwa Forest Reserve, slopes of Mnyera Peak, 2100–2200 m, 30 July 1972) differs in having shortly acuminate leaves (although some are rounded) and prominently reticulate raised venation, at least in the dry state.

GENERAL NOTE. It is difficult to decide how much of the variation in this species is due to growing in exposed windswept habitats; only field studies would answer this.

10. S. sp. A

Tree to 15 m; young branches ± 4-angular, later terete. Foliage drying somewhat yellowish green. Leaves elliptic, up to 13 cm long, 7.5 cm wide, shortly sharply acuminate at apex when young but older leaves mostly damaged at apex and appearing rounded, cuneate at the base, the venation reticulate and raised on both surfaces; petiole 8–10 mm long. Inflorescences terminal and axillary forming panicles to 9 cm wide; axes 4-angular, 1.5–3 mm wide; short pseudopedicel plus calyx ± 4 mm long including broadly triangular lobes scarcely 1 mm long. Petals white, appearing ± free, 1.5 mm long. Stamens and style not fully developed, ± 2 mm long.

TANZANIA. Mogogoro District: Uluguru Mts, Bunduki, 12 Aug. 1933, *Schlieben* 4229!
DISTR. **T** 6
HAB. Stream bank in rain forest; 1300 m

NOTE. Said to be the same as *Burtt* 5236 but this specimen cannot now be found at Kew. The possibility of hybrid origin involving *S. guineense* and *S. micklethwaitii* is not very convincing. The fact that there are specimens in the W Usambaras bearing leaves not distinguishable from the latter but also much larger leaves up to 14 × 8 cm similar in shape to those of *Schlieben* 4229 makes it likely this is only a variant of *S. micklethwaitii*, but it certainly looks distinctive.

New names validated in this part

Syzygium cordatum *Krauss* subsp. **shimbaense** *Verdc.*, **subsp. nov.**, 74
Syzygium micklethwaitii *Verdc.* subsp. **subcordatum** *Verdc.*, **subsp. nov.**, 84
Syzygium micklethwaitii *Verdc.* subsp. **subcordatum** *Verdc.* var. **dryas** *Verdc.*, **var. nov.**, 84

T - #0156 - 160425 - C0 - 244/170/5 - PB - 9789026518164 - Gloss Lamination